Chickens

Second Edition

Chickens
Tending a Small-Scale Flock

Second Edition

by Sue Weaver

HOBBY
H/F
FARM
PRESS ®

An Imprint of BowTie Press®
A Division of BowTie, Inc.
Irvine, California

Lead Editor: Jennifer Calvert
Senior Editor: Amy Deputato
Associate Editor: Lindsay Hanks
Art Director: Cindy Kassebaum
Book Project Specialist: Karen Julian
Production Supervisor: Jessica Jaensch
Assistant Production Manager: Tracy Vogtman
Indexer: Melody Englund

Vice President, Chief Content Officer: June Kikuchi
Vice President, Kennel Club Books: Andrew DePrisco
BowTie Press: Jennifer Calvert, Amy Deputato,
Lindsay Hanks, Karen Julian, Jarelle S. Stein

Text Copyright © 2005, 2011 by BowTie Press®

Front Cover Photography: (main) Bonnie Sue (left inset) Faye Pini
(right inset) www.jeanmfogle.com
Back Cover Photography: chris2766/Shutterstock.com

Library of Congress Cataloging-in-Publication Data

Weaver, Sue.
  Chickens : tending a small-scale flock / Sue Weaver. -- 2nd ed.
    p. cm. -- (Hobby farms)
  Includes bibliographical references and index.
  ISBN 978-1-935484-60-8 (pbk.)
  1. Chickens. 2. Chickens--United States. I. Title.
  SF487.W378 2011
  636.5--dc22
                              2011001599

BowTie Press®
A Division of BowTie, Inc.
3 Burroughs
Irvine, California 92618

Printed and bound in China
17 16 15 14 13 12 11   2 3 4 5 6 7 8 9 10

This book is dedicated to the wonderful folks at the American Livestock Breeds Conservancy for saving our heritage fowl and to David Puthoff for introducing me to Buckeye chickens.

CONTENTS

INTRODUCTION

Why Chickens?

Seventy years ago, throughout the countryside and in cities large and small, backyard chicken coops were the norm. Chickens furnished table meat and eggs; most everyone kept at least a few hens. Years passed and attitudes shifted; small-scale chicken keeping became gauche. By the end of the twentieth century, while agri-biz egg and meat producers, immigrants, rustics, and aging hippies were keeping chickens, cultivated urban and suburbanites were not!

The times they are a-changin' once again. As our world becomes increasingly frenetic, violent, and stressful, a burgeoning number of Americans are seeking a quieter existence. "We'll move to the country," some decide. "We'll live on a small farm and commute or work from home; we'll garden . . . we'll have chickens!"

Nowadays, from Minneapolis to New Orleans, from Los Angeles to New York City and all points in-between, throngs of city dwellers and suburbanites raise and praise the chicken. A few miles farther out, more hobby farmers are apt to raise chickens than any other farmyard bird or beast. Hens are the critter du jour.

Why keep chickens? For their eggs, of course, and (for those who eat them) their healthier-than-red-meat flesh, whether strictly for your own table or for profit, as well. Chickens are easy to care for, and you needn't break the bank to buy, house, and feed them. You may also wish, like many hobby farmers, to keep livestock for fun and relaxation. Surprisingly, chickens make unique, affectionate pets. They offer a link to gentler times; they're good for the soul. It's relaxing (and fascinating) to hunker down and observe them.

This book is meant to educate and entertain rookie and chicken maven alike. Are you with me? Then let's talk chickens!

chapter ONE

Chickens 101

Domestic chickens belong to the Phasianidae family, as do quail, grouse, partridges, pheasants, turkeys, snowcocks, spurfowl, monals, peafowl, and jungle fowl. Domestic chickens are descendants of the Southeast Asian Red Jungle Fowl (*Gallus gallus,* also called *Gallus bankiva*), which emerged as a species roughly 8,000 years ago. Today Red Jungle Fowl have disappeared from most parts of Southeast Asia and the Philippines, but a genetically pure population still exists in measured numbers in the dense jungles of northeastern India. In Latin, *gallus* means "comb," and that is how chickens differ from their Phasianidae cousins. While chickens vary widely in shape and size, all have traits in common, including general physiology, behavior, and level of intelligence.

Physiology

Chickens see in color; their visual acuity is about the same as a human's. While they don't have external ears, they do have *external auditory meatuses* (ear canals) and hear quite well. Their frequency range corresponds to ours. Their smell is poorly developed, and they don't taste sweets. They do, however, easily detect salt in their diets. Other important physiological characteristics to be aware of concern the digestive tract, internal and external structure (bones, muscles, skin, feathers), and sexual characteristics.

Digestive Tract

Chickens have no teeth. Instead, whole food moves down the esophagus and into the *crop*, a highly expandable storage compartment that allows a chicken to pack away considerable amounts of food at a time. When packed, it's externally visible as a bulge at the base of the neck. Unchewed food trickles from the crop into the bird's *proventriculus* (the "true stomach"), then to the *ventriculous* (another stomach, more commonly called the *gizzard*) to be macerated and mixed with gastric juice from the

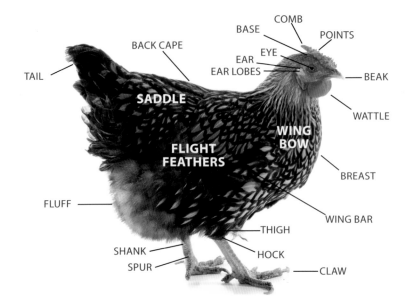

proventriculus. The food finally passes to the small intestine, where nutrients are absorbed, and then to the large intestine where water is extracted. From there it moves to the *cloaca*—the chamber inside the chicken's vent (where its digestive, excretory, and reproductive tracts meet via the fecal chamber)—and finally out the vent. Food processing time for a healthy chicken is roughly three to four hours. Urine (the white component of chicken droppings) also exits the cloaca, but via the urogenital chamber.

Chickens at a Glance

Kingdom:	Animalia
Phylum:	Chordata
Class:	Aves
Order:	Galliformes
Family:	Phasianidae
Genus:	*Gallus*
Species:	*Gallus domesticus*

Bones to Feathers

While chickens have largely lost the ability to fly, some of their bones are hollow (pneumatic) and contain air sacs. Smaller fowl can fly into trees and over fences; when harried, heavy breeds try but usually aren't able to get airborne. Chicken muscles are composed of light-color (white meat) and red (dark meat) fibers. Light muscle occurs mainly in the breast; dark muscle occurs in the chicken's legs, thighs, back, and neck. Wings contain both light and dark fibers.

Skin pigmentation varies by breed (it can be yellow, white, or black). Its exact hue is influenced by what an individual

Domestication and Cockfighting

Before chickens, there was the Red Jungle Fowl (*Gallus gallus*), a flashy, chick-enlike bird native to the forests and thickets of Southeast Asia. As a species, it emerged between 6000 and 5000 BC. By 4000 BC, *Gallus gallus* was domesticated—not for food but for cockfighting. By 3200 BC, high-caste Indian aristocrats were fighting cocks that resemble today's Aseel chickens. Chickens—and cockfighting—spread in the following centuries as traders carried domesticated birds, or chickens, farther and farther throughout the ancient world.

When Egyptian Pharaoh Thutmose III embarked on his 1464 BC Asiatic campaign, he was presented with fighting stock as tribute. The first known depiction of domestic fowl—a fighting cock—is etched on a pottery shard of that period. Cockfighting became the rage in Athens around 600 BC and was an event at early Olympic games. Greek cockfighters passed the baton to ancient Rome. An avid cocker, Julius Caesar was pleased to find the sport already established in Britain when his army invaded the island in 55 and 54 BC.

British cockfighting peaked in popularity during the seventeenth century AD. Every British town boasted a cockpit. Gentlemen breeders held tournaments, often in conjunction with horse race meets. Cockfights were held in manor house drawing rooms, schools, and even in churches. In 1792, the First French Republic took the fighting cock as it emblem; the bird also figured in the design of countless family crests and military standards in France. On the other side of the Atlantic, as well, gentlemen—including George Washington, Thomas Jefferson, and Andrew Jackson—raised and fought gamecocks; Benjamin Franklin was a noted referee.

Cockfighting remained a favored British pastime until 1849, when Queen Victoria banned it by royal decree. The sport moved underground, and the average gentleman abandoned his fighting fowl in favor of showing and creating new breeds of exhibition chickens. The movement to ban cockfighting met greater resistance in the United States. While a few progressive states outlawed it as early as 1836, most did not. When asked to support a ban, President Abraham Lincoln is said to have replied, "When two men can enter a ring and beat each other senseless, far be it for me to deny gamecocks the same privilege."

By the end of the twentieth century, the majority of American states had made the practice illegal. There were a few stubborn holdouts. According to the Humane Society of the United States, cockfighting is now illegal in every state and the District of Columbia, and any animal fighting activity that affects interstate commerce is a felony under the federal Animal Welfare Act. Thirty-nine states and the District of Columbia have made cockfighting a felony offense. Thirty-four states and the District of Columbia prohibit the possession of cocks for fighting. And forty-one states and the District of Columbia prohibit being a spectator at cockfights.

This palm-size Silkie newborn will soon lose his fluffy down. Chicks begin developing feathers within a few days of birth.

bird eats and sometimes by whether a hen is laying eggs. When a yellow-skinned hen begins laying eggs, skin on various body parts bleaches lighter in a given order (vent, eye ring, ear lobe, beak, soles of feet, shanks). When she stops, color returns in the exact reverse order.

Day-old chicks are clothed in fluffy, soft down. They begin growing true feathers within days and are fully feathered in four to six weeks. All genus *Gallus* birds, including wild jungle fowl, molt (shed their feathers and grow new ones) annually. Chickens molt from midsummer through early autumn, usually a few feathers at a time in a set sequence—head, neck, body, wings, tail—over a twelve- to sixteen-week period. Molting chickens are stressed and can be skittish, moody, and irritable. Molting hens will lay fewer eggs or stop laying altogether.

Sexual Characteristics

Growing chicks generate secondary sexual characteristics—including combs and wattles—between three and eight weeks of age, depending on their breed.

All birds in genus *Gallus*—chickens and jungle fowl—are crowned by fleshy combs and all, except the Silkie, sport a set of dangly wattles under their chins; other Phasianidae do not. Cocks develop larger and brighter-colored combs and wattles than their sisters. At about the same age, *cockerels* (young male chickens) begin crowing (pathetically at first) and sprout sickle-shaped tail feathers and pointed saddle and back feathers.

Pullets (young female chickens) reach sexual maturity and commence laying eggs at around twenty-four weeks of age. Although female embryos have two ovaries, the right ovary invariably atrophies and only the left matures. A grown hen's reproductive tract consists of a single ovary and a 2-foot oviduct or egg passageway. Her ovary houses a clump of immature yokes waiting to become eggs. As each matures—about an hour after she lays her previous production—it's

An angry hen flares her neck hackles to show she's furious. Chickens typically keep to their flocks' pecking orders, which means little infighting, but changes in routine such as relocation or flock additions can result in distressed birds.

released into her oviduct. During the next twenty-five hours, roughly, the egg inches along the oviduct, where it may be fertilized, enveloped by egg white (albumen), sheathed in a membrane, and sealed in a shell. Because each egg is laid a bit later each day and because hens don't care to lay in the evening hours, the hen eventually skips a day and begins a new cycle the following morning. All the eggs laid in a single cycle are considered a clutch.

Behavior

Chickens are easily stressed; stress seriously lowers disease resistance and stressed chickens don't thrive. Panic, rough handling, abrupt changes in routine or flock social order, crowding, extreme heat (especially combined with high humidity), and bitter cold can stress chickens of all ages. Labored breathing, diarrhea, and bizarre behavior are the hallmarks of stressed fowl. To keep stress levels low, it is important that you understand chicken behavior.

Pecking Order

In 1921, while studying the social interactions of chickens, Norwegian naturalist Thorlief Schjelderup-Ebbe coined the phrase "pecking order," now used to describe the social hierarchies of hundreds of species, including humans.

In any flock of chickens, there are birds who peck at other flock members and birds who submit to other flock members. This order creates a hierarchical chain in which each chicken has a place. The rank of the chicken is dependant upon whom it pecks at and whom it submits to. It ranks lower than those it submits to and higher than those whom it pecks at.

A flock of chicks generally has their pecking order up and running by the time they're five to seven weeks old. Pullets and cockerels maintain separate

This handsome light Brahma trio explores the outer fence line of their Missouri home on the farm of Vic and Alita Griggs.

Name That Comb

What Is a Comb?

The fleshy protuberance atop a chicken's skull is called its comb. Roosters' combs are larger than those of same-breed hens. The American Poultry Association recognizes eight basic types:

1. Buttercup: Cup-shaped with evenly spaced points surrounding the rim.
2. Cushion: Low, compact, and smooth, with no spikes.
3. Pea: Medium-low with three lengthwise ridges. The center ridge is slightly higher than the ones that flank it.
4. Rose: Solid, broad, nearly flat on top, low, fleshy, and ending in a spike. The top of the comb is dotted with small protuberances.
5. Single: Thin and fleshy, with four or five points; it extends from the beak to the full length of the head.
6. Strawberry: Compact and egg-shaped, with the larger part toward the front of the skull and the rear part no farther than its midpoint.
7. V-shaped: Two hornlike sections connected at their base.
8. Walnut: Resembles half of a walnut.

Did You Know?

- The genus name for chickenlike fowl is *Gallus*, which means "comb."

- The single comb is by far the most common type of domestic chicken comb. Its major drawback is that the points tend to freeze off in sub-zero temperatures. This doesn't affect the wearer's health, but it looks unsightly. Insulating a rooster's comb with a layer of petroleum jelly during extremely cold weather usually prevents freezing.

- The large single combs of certain breeds of hen flop over in a jaunty manner instead of standing up like those of roosters.

- To qualify them for the show ring, owners of game breeds cut off their roosters' combs and wattles in a process called *dubbing*. As game breeds were originally used for cockfighting, combs and wattles were snipped off so that an opponent couldn't grab them during a fight.

- Chickens recognize some colors and are attracted to red combs. However, not all chickens have red combs. Silkies, Sumatras, and several color varieties of game fowl have purple combs, and Sebrights' combs are deep reddish-purple.

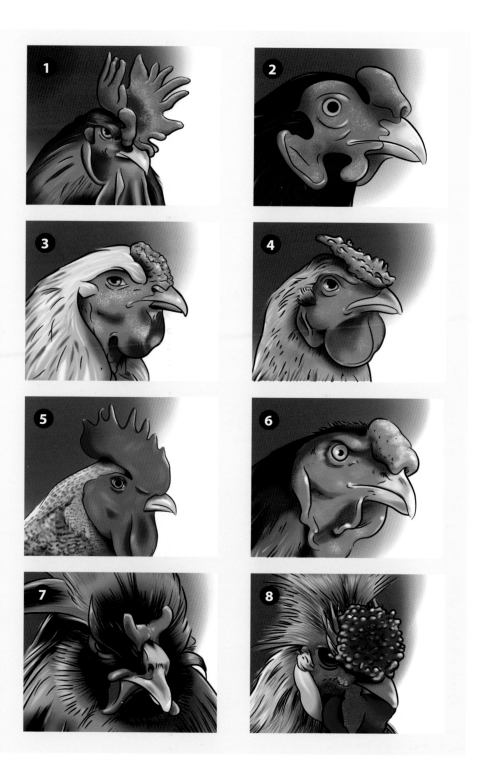

pecking orders within the same flock, as do hens and adult roosters. Hens automatically accept higher-ranking roosters as superiors, but dominant hens give low-ranking cocks and uppity young cockerels a very hard time.

In a closed flock with an established pecking order, there is very little infighting. Each chicken knows his or her place, and except among some roosters, there is surprisingly little jockeying for position. Dominant chickens signal their superiority by raising their heads and tails and glaring at subordinates, who submit by crouching, tilting their heads to one side, and gazing away—or beating a hasty retreat.

The addition of a single newcomer or removal of a high-ranking cock or hen upsets the hierarchy and a great deal of mayhem erupts until a new pecking order evolves. Because brawls are invariably stressful, it's unwise to move birds from coop to coop.

Low-ranking chickens are shushed away from feed and water by bossier birds, and they rarely grow or lay as well as the rest. Indeed, low-ranking individuals sometimes starve. If pecked by their betters until they bleed, they may be cannibalized by the rest of the flock. It's important to provide enough floor space, feeders, and waterers so that underlings can avoid the kingpins and survive.

Mating

Like adult roosters, cockerels soon begin strutting, ruffling feathers, and pecking

This red hen displays a nearly denuded back, an injury received from a rooster's sharp, spurred feet during breeding.

the earth to draw the eyes of nearby hens. This behavior is called *displaying*. Chicken mating behavior is direct and to the point. The rooster chases the hen or pullet; she crouches when the rooster mounts; insemination occurs. Cocks tread with their sharp toenails and sometimes rake hens with their spurs while mating, occasionally to the point of shredding the poor biddies' backs.

Tidying Themselves Up

When their surroundings permit, chickens are tidy birds. They preen by distributing oil (from a gland located just in front of their tails) over and between their feathers. They also dust bathe. After scratching a shallow depression in suitable earth, they lie in it and kick loose dirt over their bodies, using their feet. A shake of the feathers after dust bathing sets things right.

Chicken IQ

When they founded their show-biz animal-training business in 1943, operant conditioning (clicker training) pioneers Marian and Keller Breland began by training chickens. Among the couple's first graduates were a chicken who pecked out a tune on a toy piano, another who tap danced wearing a costume and shoes, and a third who laid wooden eggs to order (any number up to six) from a special nesting box. All were barnyard chickens rescued from a neighbor's stew pot.

With her second husband, Bob Bailey, Marian Breland-Bailey used chickens to teach animal-training courses at their Animal Behavior Enterprises in Hot Springs, Arkansas. The couple chose chickens because the birds learn quickly, energetically perform for food, and move at lightning speed. They can easily learn

Biological Makeup

Temperature: 103–103.5 degrees Fahrenheit

Pulse:
  Roosters:
  240–285 beats per minute

  Hens:
  310–355 beats per minute

Respiration:
  Roosters:
  15–20 breaths per minute

  Hens:
  20–35 breaths per minute

Chromosome count: 78

Blood volume:
Roughly 6 percent of body weight

Adult body weight:
1–10 pounds

Natural life span:
10–15 years
(pet chickens
have lived for more
than 20)

A close kin of the chicken, this Royal Palm turkey is a good choice for small farms; the breed is good for meat production for a family and for pest control on the farm.

new routines when trainees make mistakes and teach the wrong protocols. Budding trainers honed their reaction times to match those of the Baileys' chickens.

When asked by Animal Action (an Ottawa-based animal rights group) whether he considered chickens and turkeys dumb animals, Ian Duncan, professor of poultry ethnology at the University of Guelph, replied,

> Not at all....Turkeys, for example,...possess marked intelligence.

This is revealed by such behavior indices as their complex social relationships, and their many different methods of communicating with each other, both visual and vocal. Chickens, as well, are far more intelligent than generally regarded, and possess underestimated cognitive complexity.

Chickens are as intelligent as some primates, says Chris Evans, animal behaviorist and professor at Macquarie University in Australia. Chickens understand that recently concealed objects still exist, he explains—a concept that even human toddlers can't grasp. Chickens have good memories. "They recognize more than one hundred other chickens and remember them," says Dr. Joy Mench, director of the University of California-Davis Center for Animal Welfare.

Dumb clucks? No, indeed!

Chicken Classifications

Early on, the American Poultry Association (APA) devised a system for classifying chickens by breed, variety, class, and sometimes strain. A *breed* is a group of birds sharing common physical features such as shape, skin color, number of toes, feathered or non-feathered shanks, and ancestry. A *variety* is a group within a breed that shares minor differences, such as color, comb type, the presence of a feather beard or muff, and so on. A *class* is a collection of breeds that originates in the same geographic region.

The APA currently recognizes twelve classes: (Large Breed) American, Asiatic, English, Mediterranean, Continental, All Other Standard Breeds; (Bantam) Modern Game, Game, Single Comb Clean Legged, Rose Comb Clean Legged, All Other Clean Legged, and Feather Legged. A *strain*, when present, is a group within a variety that has been developed by a breeder or organization for a specific purpose, such as improved rapid weight gain and prolific egg production. Chickens may also be classified as light or heavy breeds or as layers, meat, dual-purpose, or ornamental fowl.

Farmer Vic Griggs confines this flock of young light Brahmas to a grower coop. Brahmas are among the few breeds originally recognized in 1874 by the American Poultry Association.

A Matter of Breeding

Some of today's purebred fowl (chickens whose parents are of the same breed), such as the gamecock breeds, trace their roots to the distant mists of antiquity.

Egypt's elegant Fayoumi dates to before the birth of Christ. Stubby-legged, five-toed Dorkings came to Britain with the Romans. Squirrel-tailed Japanese Chabo bantams, a miniature chicken weighing between one and three pounds, emerged in the seventh century AD. Dutch Barnevelders were developed in the 1200s, about the time Venetian merchant Marco Polo wrote of the "fur covered hens" (Silkies) of Cathay. Another Dutch chicken, the deceptively named Hamburg, has existed since the late 1600s and is likely far older than that. The crested fancy fowl we call the Polish was developed even earlier, and France's v-combed La Fletche dates to 1660 AD. Naked Necks, also called Turkens (possibly the weirdest-looking chicken of them all), originated in Transylvania before the 1700s. The first all-American fowl, the

A Dominique pullet appreciates her tasty, new pasture. This first all-American poultry breed first appeared in the 1800s, a time of increasing interest in improving and refining poultry.

Dominique, is an early nineteenth-century New England utility fowl.

However, most breeds emerged between 1850 and 1925. Although Queen Victoria's cockfighting ban had already spurred interest in exhibiting chickens, the arrival of the first Asiatic breeds set Britain afire. When Cochins were exhibited at the Birmingham poultry show in 1850, author Lewis Wright gushed, "Every visitor went home to tell of these wonderful fowls, which were big as ostriches and roared like lions, while as gentle as lambs; which could be kept anywhere, even in a garret, and took to petting like tame cats." In 1865, the Poultry Club of Great Britain produced the inaugural edition of its comprehensive *Standard of Excellence*.

America's first major poultry exposition was held on November 14, 1849. More than 2,000 birds were shown by 219 exhibitors, and more than 10,000 spectators attended. An American *Standard of Excellence* followed in 1874.

Centuries of fine-tuning breeding practices have resulted in beautifully colored birds, such as this rooster found nibbling in the grass.

When the APA published its first *Standard of Perfection* in 1874, the only chickens recognized were Barred Plymouth Rocks, light and dark Brahmas, all of the Cochins and Dorkings, a quartet of Single-Comb Leghorns (dark brown, light brown, white, and black), Spanish, Blue Andalusians, all of the Hamburgs, four varieties of Polish (white crested black, nonbearded golden, nonbearded silver, and nonbearded white), Mottled Houdans, Crevecoeurs, La Fleches, all of the modern games, Sultans, Frizzles, and Japanese bantams. In the volume's latest edition, 113 breeds and more than 350 combinations of breeds and varieties are described.

Bantams are one-fifth to one-quarter the size of regular chickens. They come in sized-down versions of most large fowl breeds, although they aren't scale miniatures: their heads, wings, tails, and feather sizes are disproportionally larger than those of their full-size brethren. A few bantam breeds have no full-size counterparts. Besides being cute, bantams can be shown, they make charming pets, and their eggs and bodies—small as they are—make mighty fine eating. The American Poultry Association issues a standard for bantams, as does the American Bantam Association. (These standards don't always agree.)

chapter TWO

Which Chickens Are Best for You?

All chickens aren't created equal. It's important to pick the ones who will meet your needs. There are countless varieties and hundreds of breeds from which to choose. With the passage of time, humans have designed chickens to fulfill every niche: cold-hardy chickens, heat-resistant chickens, chickens that don't mind being penned up. We haven't designed the perfect chicken—yet! All breeds have certain failings. Furthermore, a breed that would be a bad choice for one chicken keeper (such as hens meant to be confined who can fly out of enclosures) would be perfect for another (as free-ranging chickens, those flying hens would be able to evade dogs).

Before you can settle on the kind of chickens to buy, you need to determine what purpose they'll serve and what environment they'll live in. Do you want them for their eggs? Sunday dinner? Feathery companionship? Will they spend most of their time inside or out? Will they have to contend with sweltering summer days or frigid winter nights? All of these factors make a difference in your choice of breed.

Next you must decide whether you want day-old chicks or full-grown birds as well as how many of them to get. What advantages are there to buying a pullet rather than a chick? Is it better to start with a small flock? If you haven't already done so, you should also find out what zoning laws may apply to your keeping chickens and how they affect your decision. Do you need quieter birds?

Ask yourself the following types of questions:

- Will your birds be sequestered in a chicken house, or do you favor free-range hens? Certain breeds don't like being confined while others know nothing but. A cramped coop of ornery Sumatras is a disaster waiting to happen, and find-your-own-feed Cochins might starve.
- How much room do you have to devote to chickens? A few banties can thrive in a doghouse. A dozen 10-pound Jersey Giants? They'll need a heap more space.
- Are your neighbors close by? Squawking, kinetic, freedom-craving, fence-flying

breeds likely won't do. This is especially true if you live in the city or suburbs. We'll talk more about city chickens in chapter 3.

• Are there toddlers in your family? Testy roosters of certain breeds can injure an unwary tot.

• Do winter temperatures plummet below zero where you live? Roos with huge single combs frostbite easily and some breeds simply won't thrive in this type of weather.

• Are you in a region with hot temperatures? Fiery summer heat wilts heavy, soft-feathered breeds such as Cochins, Australorps, and Orpingtons, while other breeds take heat more in stride.

• Can you keep your top-knotted, feather-legged friends confined when the weather turns foul? Mud, slush, and fancy-feathered fowl usually don't mix.

• Would you like to preserve a smidge of living history and raise old-fashioned or endangered breeds? We'll discuss Heritage chickens in chapter 4.

• Finally, if your chicken is a pet, will you keep it outdoors with the rest of the chickens or as a household pet? A chicken in the house? Yes, indeed! Read about house chickens in chapter 11.

Though we can't tell you exactly which breed to buy—describing all the possibilities is beyond the scope of this book—we can offer general advice and name birds that will meet certain criteria. (See "Which Breed?" in this chapter.) We also list other sources in the back of the book to help you make your decision.

Leghorns are great egg-layers, but you'll need a covered enclosure if you want one because they're among the flying breeds.

In a typical strutting fashion, this silver-laced Wyandotte rooster makes his dominance known, even on the outskirts of his pasture. As both egg layers and meat producers, dual-purpose breeds (such as the Wyandotte) are great additions to any hobby farm.

Chickens for Eggs or Meat

Birds with the greatest egg-laying capacity are not the same as those who plump up into the best candidates for the local chicken fry. Still different are the chickens that are the best choice for providing both eggs and meat.

Avian Egg Machines

If you want eggs—and a whole lot of them—Mediterranean breed chickens are just your thing. Small, squawky, and hyperactive, these birds mature quickly, and then everything they eat goes into laying eggs. Undisputed queens of the nesting box are white Leghorns and hybrid layers based on this breed. Other impressive Mediterranean-class layers are the Minorca, Ancona, Buttercup, Andalusian, and Spanish White Face.

Some chickens from other classes are laying machines, too. The Campine (Belgium), Fayoumi (Egypt), Lakenvelder (German), and Hamburg (Continental Europe) are popular examples. Like their Mediterranean sisters, they tend to be flighty, specialist hens.

Meat Chickens

Meat chickens (called *broilers* or *fryers*)—usually White Cornish and White Plymouth Rock hybrids—have broad, meaty breasts and white feathers, and they mature at lightning speed. Broilers are ready for the freezer in about seven weeks, and *roasters* (which are just larger broilers) are ready in just three more.

Be aware that because they're hybrids, these birds don't breed true—meaning their chicks won't possess these stellar features. They also require careful handling; because of their abnormally wide breasts and rapid growth patterns, most become crippled as they mature.

Dual-Purpose Chickens

Dual-purpose breeds lay fewer eggs than superlayers and mature a heap slower than meat hybrids, but they're ideal all-around hobby farm birds. They're quieter, gentler, and friendlier than the specialists, and they're hardy and self-reliant, to boot. They are broody, so hens will set and hatch their own replacements. Nearly all lay brown eggs and are meaty enough to eat, should you wish to do so.

With a few notable exceptions, dual-purpose birds hail from the English and American classes. There are scores of interesting breeds and varieties.

Chickens as Pets

Do chickens make good pets? Absolutely! They're smart and affectionate, and a chicken costs little to maintain. You can teach your chicken to do tricks—it'll sit on your lap, and it may even sing if she likes you a lot. You don't need a lot of space to keep a chicken. It won't bark at the neighbors while you're at work. You can raise it from a peep for just a few dollars. All in all, a chicken makes a mighty fine friend. You can even take it along when you run errands; a chicken in your car turns heads!

If pets are your pleasure but you don't plan to handle them, most any sort of fowl will do. If you want pet chickens who are tame, that's another proposition.

Some breeds are rowdy, antisocial, and just not much fun to have around; others are downright cuddly. You want to choose pets from the latter group. Silkies, Cochins, Brahmas, Naked Necks, and Belgian D'Uccles, for example, are easy to tame and make quiet, affectionate, companion chickens. Flighty Leghorns and their ilk can be tamed—but it takes a lot more time and effort.

If you'd like eggs from your pets, that narrows the equation. Not all hens lay scads of eggs. However, most young hens

Making a Chicken a Pet

When you brood your next batch of chicks, pick one to hand tame. Carefully pull her out of the brooder for short periods every day. Cup her between your hands, and hold her near your face. Speak gently for a minute or two, then put her back. If you work with her, she'll bond with you. By the time she leaves the brooder, she'll be *your* chick.

To domesticate an older bird, work quietly and carefully. Hold her securely, so she can't flop. Stroke her wattles—chickens like that—and offer her goodies, such as bits of fruit or veggies. It won't be long until she's tame!

You might consider training your chicken as a therapy animal; hospitalized children and nursing home residents love to hold chickens. Investigate clicker training, otherwise known as operant conditioning. Widely used to train sea mammals, dogs, and horses, the techniques were originally perfected using chickens!

of the generally calm and amiable old-fashioned, dual-purpose breeds crank out one hundred to two hundred (or more) tasty brown cackleberries (eggs) a year. If a rooster fertilizes their eggs

It isn't difficult to warm up to a cute little chick. But to get her to warm up to you, spend plenty of time gently handling her and speaking sweetly to her.

and you allow it, most dual-purpose biddies will hatch chicks. Some ornamental breeds are friendly and lay well, too. But avoid flighty, sometimes pugnacious hybrid super layers and breeds from the Mediterranean class. They don't want to be your friend; they just want to lay eggs. Choose something a tad more laid back.

Big Bird or Chicken Little?

Once you've chosen a breed, you'll have to decide: chicks or full-grown birds? In most cases, the correct answer is chicks. Besides getting the most for your fowl-shopping dollar, you'll know exactly how old they are, and when purchased from reliable sources, chicks are nearly always healthy.

The Little Guys

Order day-old chicks from commercial or specialty hatcheries. The former sell dozens, sometimes hundreds, of breeds and varieties of quality chicks at modest prices. For most of us, this is the logical way to fly. Specialty hatcheries are run by knowledgeable poultry aficionados who specialize in specific sorts of fowl. You'll pay more at a specialty hatchery, but if you want to show chickens or to one day breed show-quality fowl, paying extra for specialty hatchery chicks is the way to go.

A newly hatched chick can live three days without food and water, subsisting solely on nutrients absorbed from its egg. Therefore, you can purchase chicks from hatcheries on the other side of the country, and—shipped overnight air—they should arrive safely at your nearest post office without a hitch. However, sometimes a chick does die in transit. Thus it's wise to order from the closest responsible source, so that your chicks needn't travel farther than necessary. Some hatcheries

Which Breed?

Breeds most likely to make great pets:
Barnvelder, Belgian d'Uccle, Cochin, Dorking, Jersey Giant, Naked Neck (also called a Turken), Orpington, Polish, Plymouth Rock, Silkie, Sussex

Other easygoing, friendly breeds:
Ameraucana, Araucana (usually), Aseel (cocks are aggressive toward one another), Brahma, Dominique, Faverolles, Java, Langshan, Sultan, Welsumer, Wyandotte (usually)

Cold-hardy breeds:
Araucana, Ameraucana, Aseel, Australorp, Brahama, Buckeye, Chantecler, Cochin, Dominique, Faverolles, Hamburg, Java, Jersey Giant, Langshan, Old English Game (dubbed), Orpington, Rosecomb, Silkie, Sussex, Welsumer, Wyandotte

Breeds prone to frostbitten combs:
Andalusian, Campine, Dorking, Leghorn, New Hampshire Red, New Hampshire White, Rhode Island Red. (Roosters are more likely than hens to suffer frostbite; their combs are larger, and they don't tuck their heads under their wings while sleeping as hens do.)

Silkie

Heat-tolerant breeds:
Andalusian, Aseel, Brahma, Buttercup, Cubalaya, Fayoumi, Leghorn, Minorca, Modern Game, New Hampshire Red, Rhode Island Red, Rosecomb, Silkie, Spanish White Faced, Sumatra

Flying breeds:
Ancona, Andalusian, Campine, Fayoumi, Hamburg, Lavenvelder, Leghorn, Rosecomb, Sebright, nearly all bantams

Noisy breeds:
Andalusian, Cornish, Cubalaya, Leghorn, Modern Game, Old English Game

Flighty breeds:
Ancona, Andalusian, Buttercup, Fayoumi, Hamburg, La Fleche, Lakenvelder, Leghorn, Minorca, Sebright, Spanish White Faced

Aggressive breeds:
Ancona, Aseel (cocks; toward one another), Old English Game, Cornish (cocks), Rhode Island Red (cocks), Cubalaya, Modern Game, Rhode Island Red (some strains), Sumatra, Wyandotte (some strains)

Self-reliant breeds (good foragers, ideal free-range chickens):
Andalusian, Australorp, Belgian d'Uccle, Buckeye, Buttercup, Campine, Chanteclar, Dominique, Fayoumi, Hamburg, Houdan, Java, La Fleche, Lakenvelder, Marans, Minorca, New Hampshire Red, Old English Game, Orpington, Plymouth Rock, Rosecomb, Sebright, Silkie, Sussex, Naked Neck (also called a Turken), Welsumer, Wyandotte (Avoid all-white individuals; they're more easily spotted by predators than colored and patterned varieties of the same breeds.)

Breeds that tolerate confinement reasonably well:
Araucana, Ameraucana, Australorp, Barnvelder, Brahma, Buckeye, Cochin, Cornish, Crevecoeur, Dominique, Dorking, Faverolles, Houdan, Java, Jersey Giant, La Fleche, Lakenvelder, Langshan, Leghorn, Naked Neck (also called a Turken), New Hampsire Red, Orpington, Plymouth Rock, Polish, Rhode Island Red, Silkie, Sultan, Sussex, Welsumer, Wyandotte

Breeds that don't tolerate confinement well:
Ancona, Andalusian, Buttercup, Cubalaya, Fayoumi, Hamburg, Malay, Minorca, Modern Game, Old English Game, Spanish White Faced, Sumatra

will replace chickens that are dead on arrival, but others won't. Read the guarantee before ordering chicks from a particular place. If the service is available, pay to have your chicks vaccinated for Marek's Disease. This can only be done when they're newly hatched, meaning it's now or never, and it's better to be safe than sad.

Be aware that you can't mail order five or six chicks. For the birds to stay warm enough in transit, a certain number of bodies must be in the shipping box, generating heat. It generally takes about twenty-five large fowl chicks or twenty-five to thirty-five bantams to do the trick. Some hatcheries allow you to order Guinea keets or other similar-size hatchlings to fill the quota. You can also find others interested in buying a few chicks and place a co-op order that will be shipped to one address.

If you don't want to deal with roosters, buy sexed pullets. Straight-run chicks (an equal mixture of males and females) are cheaper, but at least half will be cockerels. If you can raise and butcher the excess roosters, fine. Otherwise, buy just two or three sexed roos to add to the mix—or buy none at all. Hens don't need roosters to lay eggs.

Before your chicks arrive, assemble everything you'll need to feed, water, and brood them (keep them warm inside a heated enclosure). Have the brooder box ready and waiting. We'll talk more about this in chapter 7.

Plan to be home the day your chicks are scheduled to arrive. In most cases, they won't be delivered to your door; someone from the post office will call you to pick them up. When you arrive for the delivery, open the box of chicks in the presence of a postal worker who can verify your claim should any of them be dead. Then rush your new birds straight home to a cozy brooder box, water,

These chicks have just arrived from the feed store, safely packed in a sturdy, paper towel–lined box. Nowadays, a few hatcheries will ship as few as five or ten chicks to some locations, knowing that city chicken keepers don't need that many peeps.

This chick is suffering from pasty butt, a condition where dried droppings prevent him from eliminating.

and feed. Don't take side trips with your chicks in tow.

When you get them home, remove the chicks from their shipping box one by one and examine them. If a chick has *pasty butt* (an affliction where crusty, dried droppings block a chick's vent, making it impossible for the bird to eliminate), gently wash its little behind with a soft cloth dampened in warm water. This problem is common with mail-order chicks, especially in their first five or six days after arrival.

Check the toes. When caught early, crooked or curled toes can be splinted using wooden match sticks and strips of adhesive bandage snipped to size (see page 104). Some straighten, some don't, but you won't know unless you try! If a chick looks normal, dip its beak in water so the chick knows where the water is and starts drinking, and then place the bird gently under the heat source.

Feed stores frequently offer day-old chicks for sale. Breed selection may be limited (mail-order chicks may be your only option if you've decided to buy a rare or unusual breed) and feed store chicks aren't often sexed. However, you can choose the ones you want, buy just a few, and get them home quickly. Select bright-eyed, active chicks with straight shanks, toes, and beaks as well as clean, unobstructed bottoms. Don't buy problems—it's best to avoid chickies with issues.

The Big Guys

If you don't want to deal with tiny chicks and you're lucky, you might be able to buy sixteen- to twenty-two-week-old, almost-ready-to-lay females called *started pullets*. Initially, they cost more per bird, but you won't have the expense of brooding them and feeding them for months, so they can actually be a great buy.

Nowadays, pet shops often carry grown chickens. Breeders will sometimes part with a few hens or a breeding trio (a cock and two hens), and you can often find chickens for sale at country flea markets, poultry swap meets, or via classified and bulletin board ads. However, buying adult chickens can be risky. Not all sellers are scrupulously honest and it's easy to buy someone else's problem hens.

The waxy appearance of Dumuzi's comb and wattle is a sign of a healthy chicken. Discolored or swollen areas indicate possible maladies such as influenza or tuberculosis. Disease can spread through a flock in just a few days.

The Chicken Carry

When you go someplace to buy full-grown chickens, go prepared! Unless they've been hand-tamed, they won't sit quietly in your lap on the way home. Airline-style plastic dog crates make good chicken limousines.

Another option is a sturdy, lidded cardboard box punched with holes. If you do put a chicken in your lap, bring a towel to cover it, and wear long sleeves because scared chickens scratch.

Another thing beginners may not know: don't carry chickens by their legs! It can hurt them, it's undignified, and it scares them silly. People think chickens don't mind this position because they don't flop. Well, can you say, "shock?"

You should carry a chicken close to your body with your right arm hugging his body against yours. Then, your right hand can hold his feet while your left hand can support his chest. Give your chickens a break! How would you like to be carried around upside down?

—Marci Roberts, Springfield, MO

Ideally, you should only buy fowl from flocks enrolled in the USDA's National Poultry Improvement Plan (NPIP). These birds are certified free of pullorum (a severe, diarrheal disease) and typhoid and are healthier than your run-of-the-mill chickens. Barring these issues, choose active, alert, clear-eyed chickens with smooth, glossy feathers and bright, fleshy, waxy combs and wattles. Refuse birds that cough, wheeze, or have discharge or diarrhea. Tip the chicken forward and scope out the area around its vent, and also check under its wings. If you spy insects or eggs and you don't want to deal with parasites, you'd best not buy the bird.

If you want eggs or plan on eating the chickens, you must buy young ones. Young adults have smooth shanks; older birds' shanks are dry and scaly, and their skin is thick and tough. Cockerels have wee nubs where their spurs will grow and some pullets have them, too; long spurs denote an older bird. Press on the chicken's breastbone; a youngster's is flexible, while old chickens have rigid breastbones.

How Many Chickens?

The answer: a resounding "it depends." If you're new at chicken keeping, don't overextend yourself. Start small and learn as you go. The downside to this advice is that adding new birds to an established flock upsets its pecking order and spawns stress. Overall, it's better for your birds to hatch out a new hierarchy than for you to bite off more than you can chew.

By the same token, if you're experienced or you're certain about how many you want to keep, you'll save your chickens a lot of stressful infighting—and possibly disease—by buying all the birds you need up front and then maintaining a closed flock until you start back at square one again.

Unless you can spend a lot of quality time with a pet chicken (as you might with a house chicken), buy at least two. Chickens are sociable birds; a solitary cock or hen will be lonely.

You should also buy at least one layer hen per family member, more if your family eats lots of eggs or if you choose a dual-purpose breed. If you plan to maintain a closed flock (meaning you don't add new adult chickens to an established flock—a wise decision), you should allow for several years' flock mortality. To do this, purchase 10–20 percent more chicks than you initially think you'll need. Don't buy more birds than you can properly house. We'll talk more about this in chapter 5.

Make sure your chicken has someone to hang out with, whether that's you or a feathered friend.

chapter THREE

City Chicks

I f you live in the suburbs or in a city, you can still probably keep a few hens. True, there are limits and stipulations, and some of them are strict, but chicken fanciers across the land in cities as diverse as New York City, Los Angeles, Minneapolis, St. Louis, Seattle, Des Moines, and Fort Worth keep city chicks. Chances are, your city or town allows them, too.

Reasons to Raise City Chickens

If you're looking for a way to justify raising chickens, look no further. Here are seven very good reasons to raise city chickens.

1. Bring a sense of country to the city and a touch of the past to our busy, modern lives. Slow down. Watch your chickens scratching in the dirt, loping after bugs, being chickens. Kick back and dream of less hectic times. There, now, doesn't that feel better?

2. The eggs! The yummy, fresh-from-the-hen eggs, with yokes so rich that your mouth will water in anticipation as you conjure up that indescribable flavor. Eggs to bake with. Eggs to share with friends. And you don't need a huge flock to supply them. Three or four young hens keep the average family supplied with tasty cackleberries, often with some to spare. And there's a bonus: research indicates that chickens allowed to roam freely and nosh on grass and bugs lay eggs that are higher in omega-3 fatty acids and vitamin E. They're lower in cholesterol than commercial eggs, too.

3. What to do with the leftover salad or last night's broccoli with cheese? It's bonus feed for the chickens, of course! Chickens safely and happily devour most anything except large portions of meat or fat; raw potatoes, potato peelings, and potato vines; tomato vines; avocados, guacamole, or avocado skins and pits; tobacco (pick up those cigarette butts); and spoiled or excessively salty or sugary foods, especially chocolate. Avoid onions and garlic, as their lingering flavor

Chicken Sayings

The world's love of chickens is obvious when you think about how many of our everyday phrases involve them.

- Don't count your chickens before they're hatched.
- Don't put all your eggs in one basket.
- You have to break some eggs to make an omelet.
- Happy as a rooster in a hen house
- Fussing like a hen with one chick
- *La ley del gallinero* (Spanish—translates to "the law of the chicken house," meaning that the hen that roosts at the top poops on the ones below her)
- Madder than a wet hen
- Faster than a chicken on a June bug
- To chicken out
- Get up (or go to bed) with the chickens
- Hen-pecked
- Fly the coop
- As scarce as hen's teeth
- Cock 'o the walk
- Dumb cluck
- Rule the roost
- Cock sure
- Bad egg
- Run around like a chicken with its head cut off
- Chicken feed (a small amount of money)
- [Something is] not everything it's cracked up to be
- Something to crow about

taints those yummy eggs, and citrus or citrus peels because they tend to lower egg production. Why put food down the garbage disposal? Feed it to the chickens to make more eggs.

4. Chicken manure in your yard and garden? Oh, yes! Chicken droppings are high in nitrogen and make excellent natural fertilizer. Your gardener friends and neighbors will stand in line for your chicken-coop cleanings, and if you let your hens wander around the backyard for a few hours a day, they will help green your lawn.

5. Chickens rid your yard of bugs. Cockroaches, aphids—they're fair game to a hungry chicken, and as she scratches around, seeking tasty bugs, she automatically aerates your soil. Chickens also eat grass and weeds, which can cut down on their feed bills. If you build a chicken tractor

Is there anything better than a breakfast made with farm-fresh eggs?

(see pages 75–78) for your urban hens, you can put them to work anywhere in your yard.

6. By keeping your hens, you're saving lives. If you aren't familiar with the conditions under which factory-farmed hens are kept, research the subject now. By keeping your own infinitely healthier and happier layers, you're reducing the demand for store-bought eggs. Less demand means fewer cruelly imprisoned commercial hens. Simple.

7. And, finally, chickens are funny, low-maintenance pets. They're surprisingly intelligent and have quirky, endearing personalities. Tame chickens like to be picked up and cuddled. What's not to like about urban chickens?

First Things First: Check Those Statutes

Before you scope out a place for your coop, before you pick out hens, you must carefully—and I mean *carefully*—investigate your municipality's laws and regulations regarding chicken-keeping. Your neighbor down the block having hens doesn't mean that you can have them, too. Perhaps he's illegally keeping chickens or has applied for a personal zoning variance. The lay of his property might allow compliance to setback regulations that yours doesn't.

Many cities publish municipal statutes online. If yours doesn't, take a trip to city hall and ask to be directed to the office that oversees municipal laws. Request a printed copy of the appropriate statutes and then take them home and go over them with a fine-tooth comb. While most towns and cities do allow chicken keeping within city limits, a long list of stipulations may apply. You will be limited in the number of hens you can keep

This happy hen has the run of the farm. Check your city's laws before allowing yours to free-range—some ordinances may restrict such practices or even outlaw chicken keeping altogether.

(roosters are nearly always verboten) and where and how you can keep them.

Consider statutes in Duluth, Minnesota (chosen because we used to live just down the road), where urban chicken keeping was legalized in 2008.

- A license is mandatory and costs $10 per year. Persons convicted of cruelty to animals in Minnesota or in any other state may not obtain a license. A representative of the animal-control authority in Duluth must inspect chicken facilities prior to licensing. If the chickens become a nuisance, as evidenced by three violations of Duluth City Code within twelve consecutive months, the license will then be revoked.
- Chickens may be kept only at single-family dwellings as defined by Duluth City Code. No person may keep chickens within the single-family dwelling.
- The maximum number of chickens allowed is five hens. The keeping of roosters is forbidden.
- Chickens must be provided a secure, fully enclosed, well-ventilated, wind-proof structure in compliance with current zoning and building codes, allowing 1 square foot of window to 15 square feet of floor space. It must have a heat source to maintain adequate indoor temperatures

during extreme cold weather. The floor area or combination of floor area and fenced yard for keeping chickens shall be not less than 10 square feet of space per chicken.

- Fences around yard enclosures must be constructed with mesh-type material and provide overhead netting to keep chickens inside and predators out.
- Chickens must be kept in their roofed structure or attached fenced yard at all times.
- No chicken structure or fenced-yard enclosure shall be located closer than 25 feet to any residential dwelling on adjacent lots.
- All droppings must be collected on a daily basis and placed in a fireproof covered container until applied as fertilizer, composted, or transported off the premises.
- No person shall slaughter chickens within the city of Duluth.

These are typical statutes. Many allow renters and dwellers in two-family housing units to keep chickens but require them to obtain permission from their landlords and neighbors before acquiring hens. Some allow free-ranging chickens, some don't. Some municipalities, such as Duluth, disallow the keeping of house chickens (see chapter 11); if it's not specifically mentioned in your city's statutes, it's probably okay.

Once you have a copy of your city's chicken laws, keep it handy in case a neighbor complains. Go to great pains, however, to keep your neighbors mollified. We'll talk about that in just a bit.

Bantam Cochins like this one make great city chickens. Not only are they smaller than full-size chickens, but they have good temperaments.

Choosing a Breed

Nervous, squawky chickens that fly are not good prospects for city living. We've rated the better laying breeds in a chart at the end of this book; take a look before you pick your city chicks.

Don't overlook bantams. Bantams take up much less space than full-size chickens. As with full-size birds, some breeds are more suited to become urban layers than others. I suggest bantam Cochins, sometimes also called Pekins, like my birds Dumuzi and Marge, pictured on page 75. Cochins are gentle, quiet, attractive chickens and they aren't prone to flying. Bantam Cochins generate less manure (and thus less odor) than full-size breeds, and their peewee-size, lightly tinted, brown eggs are so tasty.

Sex-Links as City (and Country) Chickens

There are two basic types of hybrid sex-link chickens, red and black, though each goes by several names.

Black sex-links, usually called Black Stars, Black Rocks, or Rock Reds, are a cross between a Rhode Island Red or New Hampshire rooster and a Barred Rock hen. Both sexes hatch out black, but a cockerel will have a white dot on his head. Pullets feather out black with a hint of red on their necks; cockerels have Barred Rock-type plumage accented with a few red feathers.

Red sex-links, also known as Golden Comets, Gold Stars, or Cinnamon Queens, depending on the specific cross used to create them, are produced by a number of different crosses. White Plymouth Rock hens with the silver factor are crossed with New Hampshire roosters to produce Golden Comets. Silver Laced Wyandotte hens are crossed with New Hampshire roosters to produce Cinnamon Queens. Additional red sex-link combinations are Rhode Island White hens with Rhode Island Red roosters or Delaware hens with Rhode Island Red roosters. Cockerels hatch out white and then feather out pure white or with a hint of black feathering mixed in, depending on the cross. Pullets hatch out buff or red, depending on the cross, and they feather out buff or red with flecks of white throughout.

Both sex-link colors are calm, cold hardy, quiet, and friendly birds with an unusually efficient feed-conversion ratio. They work well in confinement or free-range situations. Hens weigh about 5 pounds each, begin laying earlier than most breeds, and lay a *lot* of brown eggs. People who keep them say they're wonderful pets. Sex-links are outstanding city chickens.

If you plan to raise your birds from chicks, you must buy sexed chicks, not straight-run packages, to get all pullets. Even then, there's a certain amount of error in sexing day-old chicks. If you get a cockerel in your box of pullets, what will you do? A very workable solution, if you want good layers but don't have your heart set on the standard breeds, is to buy hybrid sex-link chicks. Newly hatched sex-link pullets and cockerels are colored differently, so you can tell them apart. There's no room for error, and sex-links are very good hens.

A Home for Your City Chicks

Turn ahead to chapter 5 to read about coops and runs for your chickens, but keep these extra thoughts in mind.

Keeping a clean coop is essential to keeping chickens in the city or suburbs.

- Most city statutes spell out how big your coop and outdoor exercise area should be and which amenities you must provide. What they often don't stipulate—and this is vitally important—is that city chicken-keeping facilities must be *attractive*; neighbors won't be pleased if you create an eyesore in your backyard.
- Buy or build the best coop and fencing you can afford. Sturdy prefabricated units are especially appealing in urban situations because they're engineered to combine safety, convenience, and beauty. If you are going to build your own coop, collect ideas by surfing the Internet or visiting other chicken keepers, buying plans, or perusing a copy of Judy Pangman's book, *Chicken Coops: 45 Building Plans for Housing Your Flock* (see resources)—it's a good one!
- Obey your municipality's setback laws. Measure to be certain. In fact, to preserve your neighbors' good

will, build your facilities as far away from property lines as possible.
- Opt for a generously sized coop, based on the number of hens you plan to house. More space means less crowding and happier hens, along with less-concentrated waste and less smell.
- While some urban chicken keepers go for cute- or quaint-looking coops, in many cases, natural camouflage is more in order. A privacy fence or shrubs planted around your chicken facilities make them less obtrusive and also serve to help deaden sounds.
- Plan outdoor facilities for your hens. An exercise pen attached to their coop, a well-fenced backyard (providing your hens aren't flyers), or a chicken tractor all work well. Otherwise, plan to stay outdoors

This chicken coop in West Seattle is not only a great example of a clean city coop, but it is also an example of the fun you can have designing your coop.

with your hens while they free-range on bugs and grass. They need you to protect them from predators (particularly dogs and humans) and prevent them from wandering into neighbors' yards or the road.

Keep it Clean

A single full-size chicken can produce up to 50 pounds of solid waste per year. If you don't keep your facilities ultra-clean, they will smell. Nothing turns off picky neighbors faster than *eau de barnyard* wafting over the property line. While we usually advocate deep litter bedding for chickens (see chapter 5), in a city setting, it's better to pick up messes daily and completely strip and re-bed your hen house once a week.

Store waste in covered trash receptacles and find a place to dispose of it on an ongoing basis. If you garden, compost it. If you don't garden, compost it anyway and present finished compost to gardening neighbors and friends. Barring that, take waste to a farming friend in the country so that he or she can dispose of it. Don't let it accumulate, uncovered, on your property for very long.

Be prepared to deal with flies and rodents. Earth-friendly fly sprays and fly traps are the way to go. Rodents are a bigger problem but one you must face, as chicken feed is ambrosia to mice and rats. Store feed in covered metal containers. Trash cans work exceptionally well. Don't use plastic containers; rats chew through them without a twitch of a whisker.

Eradicating existing rodents is tricky. Don't use poisons that your neighbor's cat or a wandering toddler might find. Traps work well, but better yet, adopt a friendly cat or a Parson Russell or Rat Terrier that needs a good home; they are biological vermin-control at its finest.

Keeping City Chickens is a Privilege

It's important, both for you and for fellow municipal chicken keepers, to comply with chicken laws to the letter and not let your birds create a disturbance. It's also important to get along with your neighbors; their complaints could bring animal control to your front door. If enough neighbors complain on a citywide basis, your right to keep chickens could be revoked.

So consider sharing eggs with your neighbors. Be considerate; even if noisy roosters are legal, don't keep one. Invite neighborhood children to meet your hens and distribute chicken feed. Happy neighbors = happy you.

Finally, consider joining or creating a local city-chicken-keepers' association. Encourage members to teach community classes in urban chicken keeping. Take programs to schools and local events. Show naysayers that city chickens aren't the smelly, noisy barnyard fowl that they expect. Maybe they'll take up the banner and get some chickens, too.

Giving gifts of fresh eggs or of tasty treats made with them can go a long way toward neighborly affection.

chapter FOUR

Bring Back Those Old-Time Chickens

As mentioned in chapter 1, about 8,000 years ago, chickens were domesticated from the red jungle fowl, a sprightly chickenlike bird that still thrives in the wilds of Southeast Asia. Recent research suggests that multiple domestications may have occurred roughly simultaneously in South and Southeast Asia, in places such as North and South China, Thailand, Burma, and India.

Since then, hundreds of breeds and types of chickens have evolved through natural and human selection, all tailor-made to suit the needs of the people who kept them and the climates and conditions in which they lived. Now, they're disappearing from the earth at an alarming rate.

Chickens are not alone; all livestock species share the same fate. According to the United Nations' Food and Agriculture Organization (FAO), at least 1,500 of the world's estimated 6,000 livestock breeds are in imminent danger of extinction. The organization claims that the world is currently losing an average of two domestic animal breeds each week and that half of the breeds that existed in Europe in 1900 are already extinct.

Poultry breeds are especially imperiled. In 2004, the American Livestock Breeds Conservancy (ALBC) conducted a census of chicken breeds. Of the seventy-some breeds currently maintained by American poultry breeders, half are endangered and thirteen are practically extinct.

Industrialized farming fans the flames of this worldwide trend. Large corporations maintain factory-farmed chickens in controlled environments (eliminating the need for breeds adapted to various regions or climates); they control their birds' health through liberal doses of antibiotic cocktails (quashing the need for disease-resistant Heritage strains); and they feed their unfortunate victims high-protein, growth-hormone-enhanced feed so that they pump out an egg a day or reach market size in as little as six weeks. The result: a bountiful supply of cheap, essentially tasteless, hormone- and antibiotic-laced eggs or meat produced at the cost of the birds' health and well-being.

Heritage Chicken Versus Industrial Chicken*

HERITAGE CHICKEN	INDUSTRIAL CHICKEN
Heritage chickens are from parent and grandparent stock of breeds recognized by the American Poultry Association's *American Standard of Perfection*. The APA is the oldest agricultural organization in the United States.	Industrial chickens have been derived, over time, from multiple breed crossings.
Heritage chickens are akin to the open-pollinated varieties of heirloom fruits and vegetables that belong to no one and everyone. This is the source of genetic diversity.	Industrial chickens are carefully controlled proprietary genetic lines. Ten large companies produce more than 90 percent of the nation's poultry. This results in the loss of genetic diversity.
Heritage chickens are naturally mating, producing fertile eggs.	Industrial breeds of chickens are naturally mating, though artificial insemination can also be used. Industrial turkeys are completely reproduced through artificial insemination, and it is possible that chickens could end up with the same fate.
Heritage chicken has a rich chicken flavor and a firm texture without being tough.	The flesh of industrial chicken is very soft and bland. The modern chicken serves primarily as a vehicle for other flavors.
Heritage chickens are raised outdoors on green pasture. They actively forage for insects. Pasture production is the humane and appropriate way to raise Heritage Chicken.	Industrial chickens are typically raised indoors in confined settings. Their genetic makeup makes living outside on pasture difficult, and in some instances, inhumane.

* http://www.albc-usa.org/heritagechicken/heritagevsindustrial.html

HERITAGE CHICKEN	INDUSTRIAL CHICKEN
Heritage chickens have long, productive lives. Breeding hens will be productive for 5–7 years for hens and roosters for 3–5 years.	Industrial chickens are short lived, with breeding animals generally lasting only one reproductive cycle before being processed.
Heritage chickens grow at a more natural and normal rate, allowing the birds time to build healthy bodies and giving flavor and texture to the meat.	Industrial chickens have been selected for very fast growth, which has resulted in increased mortalities due to leg, cardiovascular, and respiratory difficulties, and producing softer, blander meat. A thin intestinal lining, as documented by a study done at North Carolina State University, makes the low feed and rapid growth of the industrial chicken possible. However, it also makes the birds vulnerable to infection and reduces disease resistance.
Heritage chickens take 16 weeks (112 days) or more to reach market weight.	Industrial chickens take 7 weeks (48 days) to reach market weight. "If a [person] grew as fast as a chicken, [he] would weigh 349 pounds at age two." (University of Arkansas Division of Agriculture, Cooperative Extension Service)
Heritage chickens look different. They have longer bodies and longer legs. They have more dark meat (thighs, legs, and wings).	Industrial chickens are round and short legged. They have more white breast meat.
Heritage chicken should be cooked slowly, and generally at a lower temperature. Fast cooking will make the meat dry and tough.	Industrial chicken may be quickly seared with high heat, in addition to being prepared slowly, at low temperatures.

Heritage Chickens

Fortunately, growing legions of poultry fanciers and small-scale chicken-raisers are stepping forth to reclaim our forebears' poultry breeds. This rare-breed renaissance is occurring throughout the world for numerous reasons.

Some conservators long for the mouthwatering fried chicken Grandma used to serve for Sunday dinner or for yummy, orange-yoked eggs with divine flavor. Some yearn to preserve living remnants of our distant past. Others do it in the name of biodiversity—they feel that if disease or genetic malady should strike down America's beleaguered battery hens and broilers, there must be hardy Heritage breeds ready to take up the slack. Some simply prefer breeds created for specific environments and needs, such as Buckeyes and Hollands for free-range eggs, Chanteclers for winter laying in the far North, or heat-tolerant Cubalayas for the steamy South.

Before you join them, get to know the American Livestock Breeds Conservancy (or the rare-breeds conservancy in the country where you live). The ALBC is a nonprofit membership organization devoted to the promotion and protection of more than 150 breeds of livestock and poultry. In service since 1977, it's the primary organization in the United States working to conserve rare breeds and genetic diversity in Heritage livestock. In 2009, the ALBC launched its Heritage-chicken promotion, and it's eager to provide new Heritage breed producers, large and small,

Centuries of fine-tuning breeding practices have resulted in beautiful birds, such as this Buff Orpington rooster.

What Is Heritage Chicken?

In order to support the APA in bringing the Heritage breeds back to popularity, the ALBC has created a list of criteria that chickens must meet to be called Heritage. The following is the definition set forth by the ALBC. Heritage chicken must adhere to all the following:

1. **APA Standard Breed.** Heritage chicken must be from parent and grandparent stock of breeds recognized by the American Poultry Association (APA) prior to the mid-twentieth century; whose genetic line can be traced back multiple generations; and with traits that meet the *APA Standard of Perfection* guidelines for the breed. Heritage chicken must be produced and sired by an APA Standard breed. Heritage eggs must be laid by an APA Standard breed.

2. **Naturally mating.** Heritage chicken must be reproduced and genetically maintained through natural mating. Chickens marketed as Heritage must be the result of naturally mating pairs of both grandparent and parent stock.

3. **Long, productive outdoor lifespan.** Heritage chicken must have the genetic ability to live a long, vigorous life and thrive in the rigors of pasture-based, outdoor production systems. Breeding hens should be productive for five to seven years and roosters for three to five years.

4. **Slow growth rate.** Heritage chicken must have a moderate to slow rate of growth, reaching appropriate market weight for the breed in no less than sixteen weeks. This gives the chicken time to develop strong skeletal structure and healthy organs prior to building muscle mass.

Chickens marketed as Heritage must include the variety and breed name on the label.

Terms like "heirloom," "antique," "old-fashioned," and "old timey" imply Heritage and are understood to be synonymous with the definition provided here.

Abbreviated Definition: A Heritage egg can only be produced by an American Poultry Association Standard breed. A Heritage chicken is hatched from a Heritage egg sired by an American Poultry Association Standard breed established prior to the mid-twentieth century, is slow growing, naturally mated with a long productive outdoor life.

Sussex chickens are on the ALBC's list of endangered breeds in the "Recovering" category. As you can see, this is a breed worth hanging onto.

with materials to help get started and, later, to promote and market eggs and meat from their Heritage chickens.

Endangered Breeds

If you join the ALBC, you'll receive the organization's bimonthly print newsletter and an annual directory full of contacts. ALBC breeders make up an active network of people who participate in hands-on conservation, marketing, and public education; if you are getting into chicken keeping, particularly if you plan on raising, showing, or breeding endangered breeds, they are definitely people you want to know.

How do you know which breeds fulfill these needs? Easy. Visit the American Livestock Breeds Conservancy website (www.albc-usa.org) and click on "Breed Information," then click on "Chickens" under "Poultry Breeds." This will bring you to the ALBC Conservation Priority List (CPL) for chickens, where breeds are categorized according to the following criteria as defined by the ALBC (on the website, you can click on the name of each breed to access pictures and information about it):

Critical: "Fewer than 500 breeding birds in the United States, with five or fewer primary breeding flocks (50 birds or more), and estimated global population less than 1,000." Breeds in the *Critical* category in 2010 were the Buckeye, Campine, Chantecler, Crevecouer, Holland, Modern Game, Nankin, Redcap, Russian Orloff, Spanish, Sultan, Sumatra, and Yokohama.

Threatened: "Fewer than 1,000 breeding birds in the United States, with seven or fewer primary breeding flocks, and estimated global population less than 5,000." Breeds in the "Threatened" category in 2010 were the Andalusian, Buttercup, Cubalaya, Delaware, Dorking, Faverolles, Java, Lakenvelder, Langshan, Malay, and Phoenix.

Watch: "Fewer than 5,000 breeding birds in the United States, with ten or fewer primary breeding flocks, and estimated global population less than 10,000. Also included are breeds with genetic or numerical concerns or limited geographic distribution." Breeds in the "Watch" category in 2010 CPL were the Ancona, Aseel, Brahma, Catalana, Cochin, Cornish, Dominique, Hamburg, Houdan, Jersey Giant, La Fleche, Minorca, New Hampshire, Old English Game, Polish, Rhode Island White, Sebright, and Shamo.

Recovering: "Breeds that were once listed in another category and have exceeded Watch category numbers but are still in need of monitoring." Breeds in the "Recovering" category in 2010 were the Australorp, non-industrial Leghorn, Orpington, Plymouth Rock, non-industrial Rhode Island Red, Sussex, and Wyandotte.

Study: "Breeds that are of interest but either lack definition or lack genetic or historical documentation." Breeds in the "Study" category in 2010 were the Araucana, Iowa Blue, Lamona, Manx Rumpy, and Naked Neck.

Buying Heritage Chickens

The same things we talked about in chapter 2 apply to choosing and buying Heritage-breed chickens with one exception: if you want to show your chickens or are very serious about conservation breeding, it's better to buy from a reputable breeder than from most commercial hatcheries. Old-fashioned breeds from hatcheries can be fine birds, but they aren't usually bred to exacting breed standards.

To locate established breeders, do an Internet search using your chosen breed's name and the word *breeder*, download a free PDF directory of Heritage chicken breeders from the American Livestock Breeds Conservancy website, or scope out the ALBC classifieds online. And if you haven't yet chosen a breed, download the ALBC's six-page "Guide to Rare Breeds of Chickens" PDF chart. It's free and very helpful.

The Hamburg is another beautiful breed on the ALBC's list.

chapter FIVE

Chicken Shack or Coop de Ville?

Chickens aren't choosy. Whether simple shack or luxurious villa, as long as the accommodations you provide meet their basic housing needs, your birds will be tickled pink—or feathery—with them. A coop must shelter its inhabitants from wind, rain, snow, and sun and protect them from chicken-swiping varmints. It also needs to be reasonably well lit and ventilated and roomy enough for the number of birds it houses (crowding causes scores of problems). When your chickens go inside, they should find sanitary bedding, roosts, nesting boxes, feeders, and waterers. For your flock's continuing comfort and for your sake, the coop should be easy to access and clean, as well.

The specific type of structure you need depends on many factors. Of first consideration are the breed and type of your chickens. For example, when it comes to indoor living space, laying hens, who will be around for many years, demand more space than broiler chickens, who have much shorter life expectancies. Bantams require less indoor space than ten-pound Jersey Giants. Outdoors, a 3-foot uncovered enclosure will keep Jerseys safely contained but will never do for flying bantams. If you don't provide the latter with a tall, covered run, you may find your entire flock going over the wall.

Climates and landscapes also shape your housing decisions. In northern climes, for instance, chicken abodes must be insulated to spare your birds frostbitten wattles, combs, and toes. In torrid southern locales, how to afford relief from the heat will be a major concern.

The cost of materials, time, and aesthetics should be factored in, as well. For example, chicken keepers without a lot of spare cash might decide to build a coop themselves rather than hiring a carpenter or buying a prefab unit. Of course, they will have to live with the results. While almost anyone can construct a functional coop from scratch, using scrounged materials at very little cost, the finished product may not jibe with a builder's preconceived image of the perfect poultry chateau. By contrast, other keepers may have the requisite carpentry skills but not the time to create their own chicken villas.

No matter how excited you are to get started, don't pick up that hammer—or let anyone else pick it up—until you've made sure the site is right. The *where* of coop building is as important as the form and method of it. You don't want to have to raze a half-constructed henhouse after a visitor helpfully points out that it's too close to the neighbor's fence. Many hobby farms are located in the suburbs, which means their coops are subject to municipal codes.

Not the least among the factors to consider when determining how and where to house your new flock are your own wishes. Remember why you decided to keep chickens in the first place. If watching hens peck in the yard will soothe your soul, it makes little sense to shut them away where you can't see them. If, however, chicken poop on your saddles makes your toes curl, locking young broilers in

a chicken yard so they don't invade your tack stall is a better option.

Your Coop: Basic Requirements

Access, lighting, ventilation, insulation, and flooring all need to be carefully considered as you plan your coop. Think in terms of easy access for you and your flock—but *not* for predators. You'll need to determine how to provide the right amount of lighting and ventilation without compromising the effectiveness of your insulation. Knowing which flooring material to use and which to avoid will save you from a lot of future aggravation.

Access

Your coop will need at least two doors: one for you and one or more for your birds. If your coop is low and close to the ground (a good design in northern climes, where body heat is wasted in taller structures), your door might

A homemade structure makes a fine grower coop for this flock of light Brahma pullets and cockerels. Vic Griggs of Thayer, Missouri, crafted this inexpensive enclosure using a second-hand truck topper and standard chicken wire.

This building, originally designed to house emus at the Griggs' farm, now serves as a chicken coop. When hawks keep their distance, the chickens enjoy the run of a chain-link fenced yard. When danger threatens, a covered rear pen shields them from predation.

simply be a hinged roof. With this kind of simple opening, you can easily feed and water your birds, tidy the coop, and gather eggs. If the coop is a standard, upright model, it should swing inward so chickens are less likely to escape when you open the door. Chicken doors (14 inches tall by 12 inches wide) can be cut in outer walls about 4–8 inches from the ground. Use the cutout to fashion a ramp. Affix full-width molding (for traction) every 6 inches along its inside surface, then hinge it at the bottom so the door swings out and down. Fit it with a secure latch so you can bar the door at night. If raccoons are a problem in the area, choose a fairly complex latch; if a toddler can open the lock, then a raccoon can unlock it easily.

Lighting and Ventilation

Light is essential to chickens' health and happiness. Natural lighting is better than bulbs and lamps. If you want your hens to lay year-round, you must wire your coop and install fixtures. Sliding windows work best; chickens can't roost on them when they're open. Every window must be tightly screened, even if your chickens can't fly. If predators can wriggle their way around or through those screens, they will. Don't use 1-inch chicken wire or poultry netting; you'll need ½- to ¾-inch galvanized mesh to keep wee beasties such as weasels and mink at bay. If you live in frigid winter climes, large south-side windows are a must; they admit lots of winter light and radiant heat. In general, allow at least 1 square foot of window for each 10 square feet of floor space. If you live where temperatures rarely dip below freezing, install even more windows. It's hard to let in too much light.

Extra windows also create cooling, healthful cross-ventilation when summer heat is an issue. Install the extra windows on your coop's north wall and possibly

Floor Space Requirements

The minimum amount of floor space needed per chicken depends on several factors, including bird type, the presence of indoor roosts, and the size of the outdoor run.

Free-range chickens and chickens with adequate outdoor runs and indoor roosts:
Heavy breeds: 5 sq. ft. per bird (2 sq. ft. if slaughtered before 16 weeks of age)

Light breeds: 3 sq. ft. per bird

Bantams: 2 sq. ft. per bird

Confined chickens without access to outdoor runs:
Heavy breeds: 10 sq. ft. per bird (6 sq. ft. if slaughtered before 16 weeks of age)

Light breeds: 8 sq. ft. per bird

Bantams: 5 sq. ft. per bird

the east one too. Your coop must be properly ventilated. Chickens exhale up to thirty-five times per minute, releasing vast amounts of heat, moisture, and carbon dioxide into their environment.

Their lungs won't sustain constantly breathing heavy, toxic air, so faulty coop ventilation quickly leads to respiratory distress. Where large windows (and lots of them) aren't possible, saw 6-inch circular or 2-by-6-inch rectangular ventilation openings high along one or more nonwindowed walls. Unplug these vents when extra air is needed, and close them tightly when it's frigid outside. Chickens can weather considerable heat or cold when their housing is dry and draft free, but they don't do well in smelly, damp conditions. If your nose smells ammonia as you enter or open your coop, it is not adequately ventilated. Do something immediately to fix this problem.

Insulation

To get your chickens through winters as unforgiving as those in northern Minnesota, the coop must be well insulated. If money is scarce, you can insulate only the coop's north wall and bank outside by using hay or straw bales stacked at least two deep. Another ploy: bank snow up against the coop; shovel, push, or bucket it as far up the sides as you can. Window height is good, if you can manage. If it's still too cold inside the coop, you'll need a heat lamp. But remember: fallen heat lamps can, and often do, spark fires. So install your heat lamp in a reasonably safe location and use it only when really needed.

Chickens can die in temperatures higher than 95 degrees. If sizzling, muggy summers are common in your locale, make sure your coop and outdoor enclosures are situated in partial shade—or plant vegetation around your chickens' lodgings to partially shade it; giant pumpkin or bottle-gourd vines on trellises are helpful. Insulation helps repel daytime heat, and fans generate badly needed airflow. Opt for light-colored or corrugated metal roofing and paint external surfaces a matte white color to reflect the heat. Avoid overcrowding by allowing additional space for each of your birds; overcrowding leads to higher indoor temperatures and humidity.

Flooring

Your coop's floor may be constructed of concrete, wood, or plain old dirt. Concrete is rodent-proof and easy to clean, but it's comparatively expensive. Wood must be elevated on piers or blocks; it looks nice but can be hard to clean and periodically needs replacing. Well-drained dirt floors work fine. However, if a dirt floor is poorly drained or allowed to become mucky, you'll have a sheer disaster on your hands.

For the best-bet deep bedding system, blanket your floor of choice with a cushy layer of absorbent material to keep things tidy and fresh. Chopped straw (wheat straw is best) or wood shavings are ideal; rice or peanut hulls, sawdust, dry leaves, and shredded paper work well, too. Line the floor 8–10 inches deep. After it's in, continue to remove messes and add more material when necessary.

Once or twice a year, strip everything back to floor level and start again. Deep bedding nicely insulates a chicken coop floor, it's the essence of simplicity, and it works!

Your Coop: Basic Furnishings

Basic coop furnishings include roosts, nest boxes for laying hens, feeders, and watering stations. Roosts are the elevated poles or boards on which chickens prefer to sleep. Roosting helps them feel safer at night, making these perches a must if you want happy hens. Keeping

A rustic A-frame coop on the edge of the garden makes a fine summer home for these free-ranging birds.

No-nonsense egg and meat producers may scoff, but for the rest of us, chicken toys in the coop are a nice touch. They're optional, but your birds will love such diversions. Happy chickens develop fewer stress-induced vices, and they tend to lay a lot more eggs. Poultry keepers theorize that stressed birds lay fewer eggs and are far more prone to disease and cannibalism than well-cared-for birds.

• • •

Joys and Toys

Consider diversions for the flock.

• Movable outside coops ("chicken tractors") make happy, healthy birds. Inside toys and diversions are important. Our chickens play tether-lettuce (pecking at mesh bags with greens suspended from the ceiling). Mirrors and bells are a hit. Rotate the toys.

• Our flock likes to peck at treats placed on empty margarine tubs. It sounds like rain on a tin roof when they pick up grain. They like the sound, too.

• I've heard that large flexible pipe (drain tile) is a hit if the birds can fit easily inside.

• Chickens in the wild investigate and scratch. Their brains are wired for cautiously investigating (they are a prey species) new situations. Ramps and things to duck under and fit inside of are popular.

• There needs to be somewhere for a sad, slightly sick, or not aggressive bird to sit and read a book. They need an escape route.

• Chickens love to scratch. A sand box with a little grain will give a thrill.

• They love greens. A box framed with two-by-fours with a wire mesh on top will let you grow grass, but keep your chickens above the grass so they don't destroy it; they can only get the long pieces.

• Perching birds (some Silkies, too) love varied heights—not just the same old perch.

—Alan Stanford, Whitewater, WI

Chicken A-Cord

Build your chicken house within two long heavy-duty extension cords' reach from your house or garage. Then you can have electricity when you want it and it's less costly than hiring an electrician.
— Mandy Anderson,
Anoka, MN

The Golden Rule

The Chicken Golden Rule: If it's possible for a chicken to hurt itself on something, it will find a way. This is important to re-member when building a coop because if you leave that one little piece of sharp wire poking out, somehow, some bird will find a way to impale itself on it no matter how remote the chance may seem. People should get down on their hands and knees—at chicken level—and really *look* at their finished coop to try and spot chicken hazards, just like new parents do for babies around the house—after all, having birds is like having an eternal two-year-old.
— Susan Mennealy,
Norwalk, CA

yourself happy makes nesting boxes a necessity—unless you like going on egg hunts or prefer your eggs precracked. Contentment and peace for everyone means making sure that there's more than one set of waterers and feeders and that they're well placed.

Roosts and Nesting Boxes

In the winter, roosted birds fluff their feathers and cover their toes; they tend to stay warmer that way. Roosts can be as simple as old wooden ladders propped against a chicken coop's inner walls. If you use a ladder, tack the top to the wall. Then, make sure the bottom is out far enough from the wall so birds settled on one rung don't poop on flockmates roosting on rungs below. If you build traditional stair-stepped roosts for your birds, set the bottom perch about 2 feet from the floor and set higher rungs an additional foot apart. Two-by-two boards with rounded edges make ideal roosts for full-size chickens; 1-inch rounded boards or 1-inch dowel rods are fine for bantam breeds. Tree branches of the same diam-eters make fine roost rails, too. But don't use plastic or metal perches; chickens require textured perches that their feet can easily grip. Allow 10 inches of perch space for each heavyweight chicken in the coop; provide 8 inches and 6 inches for light breeds and bantams, respectively.

Wherever you place your roosts, make certain sleeping chickens aren't perched in cross-drafts. Check frequently and move them if necessary.

Nesting Boxes

Hens instinctively seek dark, secluded spots to lay their eggs. Unless you provide nesting boxes, free-range hens sneak off to lay in nooks and crannies and you may never find them! Confined hens plop eggs out wherever they can, which can result in poop-splotched, cracked eggs. Specially designed nesting boxes with slanted tops and perches in front work best (you can buy ready-made wooden, metal, or injection-molded plastic ones from poultry supply houses). However, any sturdy cubicle with a top, a bottom, three enclosed sides, and bedding inside will do nicely.

Make sure the box is larger than the chicken and leave the top off in steamy summer climates. A 14-inch wooden cube with one open side makes an ideal nesting box for full-size hens; 12-inch cubes accommodate bantams with ease.

If you use a regular box, attach a 3- to 4-inch lip across the bottom front to keep bedding and eggs from spilling out. Mount the unit 2 feet from the floor in the darkest corner of the coop. Provide one nesting box for every four or five hens.

Feeders and Waterers

Place at least two sets of commercially made feeders and waterers in every coop and locate each set as far from the others as you can. This prevents guarding by militant high-ranking flock members, causing lowest-ranking chickens to starve. Instead of setting units on the floor, install them so the bottoms of the waterers and the top lips of the feeders are level with the smallest birds' backs. They'll stay cleaner that way, and your chickens will waste less food. Be sure to provide one standard hanging tube-style feeder per twenty-five chickens. If you prefer trough feeders, allow 4 inches of dining space per bird when deciding what size to purchase.

Outdoor Runs: Sunshine and Fresh Air

Another way to keep your chickens happy and healthy is to get them into sunshine and fresh air almost every day. To do that, you'll probably need a chicken pen (or run) attached to your coop. Chickens can free-range (wander wherever they like), but "wherever they like" may be in your garden (or in your neighbor's garden!), or in your garage or barn, or in places where stray dogs or other wild things can attack them. To save their skins—and your infinite vexation—consider providing a fenced-in area.

The Comforts of Home

Be sure to provide a shade area for your flock in the outdoor run. No trees? Stretch a tarp across one corner on steamy summer days. Hook it to the fence with bungee cords; a woven wire armature underneath will help secure the tarp in place.

Chickens enjoy lounging under outdoor shelters when it rains. So, if you leave the tarp up when it rains, use an ice pick, awl, or large nail to jab a few holes so runoff can drain. Don't poke holes near the center where your birds gather.

While you're at it, add a sand pit for dust bathing, as well!

This Dominique hen nibbles the grass through her double-fenced enclosure. The coop's outer layer of livestock paneling is more effective at keeping predators out than its flimsier inner layer of lightweight wire fencing, which is better at keeping chickens in.

Fencing Them In

Chicken runs are traditionally crafted of chicken wire (also called poultry netting) that is a flimsy, 1-inch mesh woven into a honeycomb pattern. If a dog or larger varmint wants to get at your chickens, this lightweight wire is not going to be a deterrent. If you value your birds, don't use chicken wire except for indoor applications and chicken-run ceilings. Workable alternatives include substantial posts with attached medium- to heavy-duty yard fencing or sturdy welded wire sheep panels (sometimes installed two panels high), or electroplastic poultry netting.

If predators, including dogs, are an ongoing headache, a strong electric charger and two strands of electric wire fencing can provide effective but cheap insurance. String one strand on 10-inch extension insulators 4 inches from the ground, along the outside bottom of the run. Using the same type of insulator, stretch another strand parallel with the top of your existing fence. These wires will prevent hungry critters from tunneling under your chicken-run fence or scaling its perimeters.

Allow at least 10 square feet of fenced run for each heavy chicken in your flock, 8 square feet for light breeds, and 4 square feet for bantams. In general, you can contain your chickens using a fence 4–6 feet high. However, most bantams and certain light full-size breeds can neatly sail over 6-foot barriers. Keep them in by installing netting over the enclosure.

The Location

Don't create your coop area too far from utilities, especially if you must carry water or run a hose or electrical extension cord

Our hen Gracie preferred living in the sheepfold to life among her peers. So we fastened a dog carrier 4 feet up the wall to create a nesting box that won't get bumped by the sheep.

from your house, garage, or barn. This will simply be a hassle for you if not taken into consideration. Make sure your coop area is reasonably distant from neighbors' property lines, especially when setback regulations are part of municipal or county codes.

Take into consideration the dryness of the ground when selecting a location; choose a well-drained area where storm runoff and melting snow won't make chicken coop floors and outdoor enclosures a continual, sodden mess. Remember, your coop area should be close enough for you to enjoy your chickens because that is why you wanted them in the first place.

Building a Cheaper Chicken Coop

If your desire to keep chickens exceeds your ability to buy or build a standard chicken coop, don't worry: chickens can adapt to simple accommodations. If you keep them dry, safe, and out of drafts, they'll be happy. Consider the following inexpensive options.

You can build chicken accommodations into existing structures such as your garage or back porch, the kids' abandoned playhouse, and unused sheds and storage buildings. Build coop-style quarters or house your chickens in cages. Show chicken fanciers often cage their birds in wire rabbit-style hutches. Buy them new or from a rabbitry that is going out of business, or build them yourself.

It's best to forgo keeping heavy breed chickens if they must live in cages; continually standing on wire floors will likely damage their feet. Allow 7 square feet of cage floor space for light breed chickens or 6 square feet for bantams.

Place a 2-foot-by-2-foot sheet of salvaged cardboard box stock in one corner where inhabitants can rest their feet; replace it with a new piece when it gets soiled.

Caged chickens appreciate roosts—affix them at least 6 inches from the floor—and boredom-squelching amusements such as toys.

A large wooden packing crate fitted with a hinged roof makes a dandy indoor or outdoor coop. Prop the lid open during the day (if you keep flying breeds, you'll have to fashion a screen) and close it at night. Outdoors, install a latchable dog door in one side and attach a small fenced run.

If appearance doesn't matter, fashion a funky, cost-effective walk-in coop out of tarps, a welded wire cattle panel (or two), and scrap lumber. Coops of standard designs can be crafted using secondhand lumber and other scrounged or recycled goodies. It's amazing what can be done with plastic tarps if aesthetics don't count.

Owls and Weasels and 'Possums, Oh My!

It's no fun—for you or your birds—when hungry midnight marauders visit the chicken coop. The best way to thwart potential predators is to lock your birds inside a safe, secure chicken coop at night. Another approach, if it's legal in your state, is to humanely trap and relocate bothersome nighttime marauders.

Set atop a piece of plywood cut to fit, this small chicken tractor can also serve as an inexpensive cage inside a house. Avoid cages with wire floors or install cardboard or wood over the flooring—heavy breed chickens are especially susceptible to foot injuries from long days on wire–bottom cages.

Measures for Keeping the Varmints Out

After sundown, every door, window, and any other crack or portal in any outside wall should be tightly blocked or screened. To be effective, screening must be small-holed and made of strong material. A mink or weasel can easily slip through 1-inch chicken wire, and larger species can simply rip it down. Choose ¾-inch or smaller mesh galvanized hardware cloth for screening windows and building outdoor enclosures to save your chickens' lives.

To discourage chicken-swiping predators, install concrete block foundations into coop floors, set at least two rows high. In addition, bury outdoor enclosure fencing at least 8–12 inches into the earth. Make sure the buried fencing is toed outward, away from the fence line. If winged predators, such as daytime hawks or nighttime owls, pursue your birds, cover the outdoor enclosure with chicken wire; for this purpose, it works nicely.

If your birds range freely and you live where chicken-thieving hawks wreak havoc, think camouflage. For example, don't choose a white-feathered breed. Plant ground cover—bushes, hedges, and flower beds—so that your chickens can hole up in the greenery when predators soar above.

If predators are simply getting into your feed, you can secure container lids using bungee cords. Do this by securing bungee cords to eyebolts set in the feed-room wall. The key word is *secure* for both your chickens and their feed.

Trapping Intruders

If it's legal in your state, you may decide to take care of the problem of predators by trapping and relocating them. The first step is to figure out what's killing your chickens; this will enable you to bait the live trap with species-specific yummies.

Determining Who Done It

You can figure out the identity of your midnight raider by discovering what was taken and what was left. Opossums, skunks, and raccoons savor chicken eggs. They may raid your coop without killing a bird. A dead bird, however, doesn't rule them out. They might dine on meat and eggs during the same visit.

When opossums have a hankering for chicken, they'll usually kill a single bird per visit; typically only the abdomen will be eaten. Raccoons visit a coop infrequently, once a week or so. They prefer heads and crops; more than one chicken may be killed. If skunks invade your chicken house, they're also likely chew off one or more chickens' heads; worse, a lingering aroma of *eau de phew* will generally give them away. A neat stack of dead chickens—with necks eaten and heads missing—suggests the culprits are probably minks or weasels.

Foxes usually are blamed for most chicken coop predation, but if you find dead chickens, foxes likely aren't involved. They carry their prey away with them—as do bobcats, coyotes, and predatory birds—usually without leaving a trace.

If chickens are missing and you suspect foxes, bait live traps with chicken entrails or raw chicken, dead mice, or commercial lures. These work for the wily coyote, too, but you'll need a much larger trap to catch him. Opossums like fruit, especially melon or apples; raccoons fancy marshmallows, sweet corn, canned creamed corn, and honey.

In general, baits of choice include fresh fish, chicken heads or entrails, fresh

A carefully concealed Havahart trap helped capture the critter alive so we could relocate him to a distant wildlife preserve.

raw liver, crisp bacon, fish-flavored moist cat food, and table scraps. Whatever your bait, be sure to replace it every day.

Catching the Suspect

Heavy-gauge wire live traps can be used for humane trapping and relocating. You can purchase them at most hardware stores or order them online. If you don't want to buy, contact your city or county animal control officer or humane society to see if you can get one on loan. Check out rental centers, as well; they often carry these traps in stock. If you buy a live trap, read the instructions. If you rent or borrow one, ask an experienced live trapper to teach you how to operate it efficiently and humanely.

Keep the animal's suffering to a minimum. Check traps at least three times a day (more frequently is better). When trapping during the day, make certain the traps are set in the shade. If you find an animal in the trap, immediately shroud the entire enclosure in a sheet or towel to calm the inhabitant. Promptly take it to your local animal control officer or relocate it at least five miles from your property in an uninhabited area that offers an abundance of species-specific natural food, water, and shelter.

If you catch more than one of a kind, release them in the same location. Wear gloves when transporting trapped animals; frightened wild things bite hard!

One more word of advice: you can easily live-trap and release skunks, but it takes special finesse not to get sprayed. Know what you're doing before you try.

chapter SIX

Chow for Your Hobby Farm Fowl

W hile your chickens' nutritional needs vary depending on age, sex, breed, and use, their diets must always include water, protein, vitamins, minerals, carbohydrates, and fats in adequate quantities and proper balance. All chicken keepers are in agreement on this point. But when it comes to the question of how best to supply all those dietary elements, it's a different story. Ask any group of fanciers how each feeds his or her chickens, and you'll find two distinct camps: those who never feed their birds anything except commercial mixes, and those who never feed their birds commercial mixes without supplementing them. Opinion runs high on both sides about which approach is better.

Which one you should take really depends on your primary reason for keeping chickens. If you raise birds strictly for their meat or eggs, commercial feed is the way to go. Commercial bagged rations are formulated to serve up optimal nutrition, thus creating optimal production. Supplementing commercial feed with treats, table scraps, scratch (a whole- or cracked-grain mixture chickens adore), or anything else will upset that delicate nutritional balance.

However, if like us you see your chickens as friends and don't care if their growth is slightly slower or if they produce fewer eggs, then consider supplementing their diets. They'll appreciate the variety, and you'll appreciate the much lower cost of a supplemented diet.

Water

Consider this: an egg is roughly 65 percent water, a chick 79 percent, and a mature chicken 55 to 75 percent. Blood is 90 percent water. Chickens guzzle two to three times as much water as they eat in food, depending on their size, type (layers require more water than broiler chickens), and the season—up to two or three cups per day. So whether you use a commercial or home-based diet, your chickens require free access to fresh, clean water.

Chickens need water to soften what they eat and carry it through their digestive tracts; many of the digestive and nutrient absorption processes depend on water. In addition, water cools birds internally during the hot summer months. If you eliminate water from your chickens' diet, expect problems immediately. Even a few hours of water deprivation affects egg production.

Chickens don't drink a lot at any single time, but they drink often. However, water temperature can affect how much they will drink. They don't like to drink hot or icy water, so keep waterers away from heat sources and out of the blazing sun. When temperatures soar, plop a handful of ice cubes in the reservoir every few hours. In the winter, replace regular waterers with heated ones or add a bucket-style immersion heater to a standard metal version. You can also swap iced-up waterers for fresh ones containing tepid water every few hours. In subzero climates, heated waterers are a must; even a heated dog-watering dish is acceptable.

Because chickens are inherently messy, chicken-specific waterers are better than buckets and dishes. Hanging (tube style) models are good; automatic

A large, hanging-style waterer, such as this one we use in our daytime enclosure, helps keep your birds' water free of debris.

waterers work best of all; and metal waterers last longer than plastic ones. Read the instructions that come with the waterer you choose to determine how many units you'll need.

Even if one waterer is enough, choose two. Otherwise bossy, high-ranking chickens in your flock's hierarchy may shoo underlings away from the fountain. (This recommendation goes for feed troughs, too).

Although hanging waterers can be placed on the floor, hanging them from hooks or rafters with the drinking surface level with your smallest chickens' backs will give the best results. If you can't hang a waterer, make certain it's level, or it will leak.

Whichever type of waterer you use and wherever you hang your waterers,

Waste Less

Chickens waste about 30 percent of feed in a trough feeder that's full. If the same trough is only half full, they waste only 3 percent.

Save yourself some time and money by spreading less feed for your chickens in more troughs.

clean and rinse them every day. Scour them once a week (more often in the summer, when they tend to get scummy) using a stiff brush and a solution of about nine parts water to one part chlorine bleach.

Commercial Feeds

Whether you buy or mix it yourself, a healthy, happy chicken's diet should provide the following:

- Sufficient protein based on the age and needs of the bird.
- Carbohydrates, a major source of energy.
- Thirteen vitamins to support growth, reproduction, and body maintenance: fat-soluble vitamins A, D_3, E, and K, and water-soluble vitamins B_{12}, thiamin, riboflavin, nicotinic acid, folic acid, biotin, pantothenic acid, pyridoxine, and choline.
- Macro minerals (those needed in larger quantities) and trace or micro minerals (needed only in minute amounts) to build strong bones and healthy blood cells, supporting enzyme activation and muscle function, and regulating metabolism. Hens require additional minerals, especially calcium, to lay eggs with nice, thick shells.
- Fats for energy and proper absorption of fat-soluble vitamins and as sources of fatty acids, necessary for supporting fertility and egg hatchability.
- Commercial feeds, presented at recommended levels, are designed to meet those needs precisely. To meet protein requirements, commercial feeds include a variety of high-protein meals made of corn gluten, soybeans, cottonseed, meat, bone, fish, and dried whey. Too much protein can be as bad as too little, so balancing this nutrient is especially tricky.
- Carbohydrates are much easier; they naturally compose a large portion of every grain-based diet. While some of the thirteen vitamins listed above are plentiful in natural foodstuff, commercial feeds cover all bases by

Most feed stores carry a variety of bagged chicken feeds like these. Be sure to carefully read the sacks' labels. Feed types designed for specific groups of birds may contain ingredients that your birds don't need.

CHICKEN FEED
Which commercial chicken feed should you buy?

Feed Type	Protein Content	Use It	Feed It
Starter (for layers)	18%–20%	As a high-protein feed for fast-growing future laying hens	Until 6 weeks of age
Starter (for broilers)	20%–22%	As an even higher-protein feed for even faster-growing meat chicks of both sexes	Until 6 weeks of age
Grower (for layers)	15%–16%	As a lower-protein feed for pullets	6–20 weeks or until laying begins, whichever comes first
Layer	16%–18% (with additional calcium and extra vitamins added)	As a maintenance diet for hens laying eggs for consumption	From 20 weeks on
Breeder	18% (additional calcium and extra vitamins added)	As a maintenance diet for hens laying hatching eggs	As long as hens are used for producing hatching eggs
Finisher (for broilers)	18%–20%	As a lower-protein-than-broiler-starter diet to carry broilers until slaughter	4–6 weeks until slaughter

adding a vitamin premix. Ground oyster shells or limestone, salt, and trace mineral premix are commercial feed additives designed to meet a chicken's macro and micro mineral needs. As for fats, commercial feeds contain processed meat and poultry fats in measured amounts. Fats provide twice as much energy as other feed ingredients, making them especially useful in starter feeds and growing rations.

• Mixing your own commercial-style feed is an option (and often a must for producers of organic meat or eggs), but it's a complex and nutritionally risky one. Using the commercial feed on the market is convenient and easy.

Common Ingredients and Additives

Common ingredients in commercial feeds include corn, oats, wheat, barley, sorghum, milo, soybean, and other oilseed meals, cottonseed or alfalfa meal, wheat or rice bran, and meat by-products, such as bonemeal and fishmeal. Ingredients are finely ground to produce easier-to-digest mash; sometimes, they are pelleted or processed into crumbles so there is less wasted food.

Commercial baby chick food is usually medicated; some feeds for older chickens are medicated, too. Each type, designed for a specific group of birds, contains nutrients in slightly different measures, so when buying feed, read the tags and labels to make certain you're buying what your chickens require.

Commercial feeds also contain ingredients that many fanciers don't approve of, such as antibiotics and coccidiostats for birds that don't need them, pellet binders to improve the texture of pelleted feed, and chemical antioxidants to prevent fatty ingredients spoilage. If you'd like to offer your chickens commercial feed but want to avoid the questionable additives, ask your county agricultural agent or feed store representative what "natural" commercial feeds are available locally. Some—such as Purina's Sun-Fresh Start & Grow, Layena, Flock Raiser, and Scratch Grains—can be purchased or ordered in most locales. The Murray McMurray Hatchery (see Resources) sells organic feed, priced to ship post paid to any location in the continental United States. Always be sure to read labels of unfamiliar commercial feed before dishing out the grub. Know what's in there and precisely what amounts to feed.

When feeding commercial products, choose the correct feed: starter, grower, layer, breeder, or finisher. That information will be printed on the label or tag, so always check to be certain.

Don't indiscriminately substitute any type. In a pinch, you can adjust a feed's protein level by diluting it with scratch or adding a separate protein supplement, but if the feed mill doesn't offer what you need, it's best to shop elsewhere.

Make sure to buy quality feed from a reputable source. If in doubt, major companies such as Purina and Nutrena are reliable bets.

Maintaining Nutritional Value and Freshness

To retain full nutritional value and assure freshness, purchase no more than a two- to four-week supply of commercial feed. Don't dump new product on top of remaining feed; use up the old feed first or scoop it out and place it on top of the new supply. When storing feed, place it in tightly closed containers and store in a cool, dry place out of the sun. Plastic containers work best, but if plastic-gnawing rodents are a headache, store grain in lidded metal cans. A 10-gallon garbage can—plastic or metal—can hold 50 pounds of feed and makes a neat, ready-made feed bin.

If your chickens refuse the commercial feed, examine it closely. Sniff. It may be musty or otherwise spoiled. If it seems all right, you're probably dealing with picky chickens who prefer scratch, treats, and table scraps. You should cut back on goodies until they eat the chicken ration, too. Distributing treats only in the late afternoon, after they've dined on their regular rations, will encourage them to be less picky.

The Supplement Approach

According to proponents of supplements, hens fed strictly with commercial feed lay tasteless, thin-shelled supermarket-quality eggs and broilers fed the same diet will taste like packaged, store-bought chickens. A push? Maybe. That's something you'll have to decide for yourself. What we present here are methods chicken keepers can use to supplement the diets of their flocks.

Grit and Oyster Shells

Since chickens don't have teeth, they swallow grit—tiny pebbles and other hard objects—to grind their food. If your chickens free range or you use easily digestible commercial feed, you won't need to provide your birds with grit.

Otherwise, commercial grit (ground limestone, granite, or marble) can be mixed with their scratch or container-fed to chickens on a free-choice basis.

Ground oyster shell is too soft to function as grit, but it's a terrific calcium booster for laying hens. Feeding oyster shell to hens on a free-choice basis allows the hens to eat it when they wish.

Scratch

While university resources advise a straight commercial diet, most hobbyists and small flock owners supplement this with scratch. Scratch is a mixture of two or more whole or coarsely cracked grains, such as corn, oats, wheat, milo, millet, rice, barley, and buckwheat.

Chickens adore scratch grains. Chickens instinctively scratch the earth with their sharp toenails to rake up bugs, pebbles for grit, seeds, and other natural yummies; strewn on their indoor litter or anyplace outdoors, scratch satisfies that urge. Or you can place feed in separate indoor feeders. To preserve its nutritional balance, commercial feed should be supplemented with scratch in measured proportions.

Greens and Insects

Hobby farmers and poultry enthusiasts often grow "chicken gardens" of cut-and-come-again edibles like lettuce, kale, turnip greens, and chard. Chickens of all types and sizes relish greens. Greens-chomping hens lay eggs with dark, rich yolks.

Insects add protein to chickens' diets. Free-range chickens harvest their own bugs but coop and run-caged birds don't have that chance. Capture katydids, grasshoppers, and other tasty insects to toss to your chickens. If you do, they'll soon come running when they spot you.

Good Home Cookin'

Chickens happily devour table scraps. Avoid fatty, greasy, salty stuff, anything spoiled, avocados, and uncooked potato peels. Also, strongly scented or flavored scraps, such as onions, garlic, salami, and fish, can flavor hens' eggs. Most everything else from your table will work as well—even baked goods, meat, and dairy

PROTEIN	
To increase a 15% Protein Layer Ration...	
...using a 10% protein scratch mix of these proportions...	...requires mash containing this amount of protein
half scratch/half mash	20%
two-thirds scratch/one-third mash	25%
three-quarters scratch/one-quarter mash	30%

Two of our farm favorites, Dumuzi and Marge, nosh on lettuce in the garden. Veggies are one of many treats chickens appreciate straight from your table.

products. Your chickens will love it all.

Many fanciers scramble or fry eggs and feed them back to their chickens. Egg yolk is a chicken's first food; it's a fine supplement for adult birds of all ages and an ideal use for eggs with cracked or soiled shells. However, if you overfeed your chickens on scratch, greens, veggies, and "people food," some will turn up their beaks when served commercial rations. Unless you devise a balanced diet based on home-mixed victuals, consider them supplements, not first-line chicken feed.

Many folks assume free-range chickens will grow healthy eating seeds, weeds, and bugs. They won't. However, if you supplement free-range findings with scratch or commercial feed, your chickens will cheerfully rid your yard and orchard of termites, ticks, Japanese beetles, grasshoppers, grubs, slugs, and dropped fruit. One caveat: they'll also strip your garden clean; think "fenced garden" if you raise free-range chickens.

Chicken Tractors, Pastured Poultry, and Free-Range Chickens... What Does It All Mean?

Most range (also called short-pasture) poultry systems fall into one of three basic categories: free range, pastured poultry, and chicken tractors. The first two systems are used for raising commercial flocks and probably won't interest most who read this book, but they work nicely for hobby farmers seeking ways to raise and market homegrown, value-added products such as natural or organic meat chickens and organic eggs. Chicken tractors are lightweight, bottomless shelter pens designed to move

If you have the yard space and a nicely fenced perimeter, and you know how to guard against predators, you can allow your chickens to range during the day.

wherever grass control and soil fertility are required. They appeal to small-scale raisers like you and me: maintaining a chicken tractor in your garden not only enriches your soil, it's also the perfect way to supplement your birds' diet with yummy greenery, crunchy bugs, worms, and grubs.

In the grand scheme of things, free-range chickens are any that are allowed to roam fenceless and free during the day, returning at night to a coop or shelter in a barn, tree, or wherever else they can find. Free-range chickens, when spoken of in conjunction with today's range-poultry management phenomenon, are something else indeed.

They are raised in 8-by-18-foot skid houses set in roomy, clipped grass pastures, which are surrounded by predator-resistant perimeter fencing. Each shelter has a roof, a sturdy floor, and wood-framed 1-inch chicken-wire sides.

They're fitted with roosts, nest boxes (for layers), bulk feeders, and float-valve waterers. Chickens, stocked at the rate of 400 per acre, are allowed to free range during the day and are confined to their skid houses at night. Spacing can be no less than 150 feet between skid houses or from a skid house to the nearest fence. When grazing around the skid houses grows thin (generally every few months), the skids—chickens and all—are towed

Food Is Not Boring Here!

My chickens get yummy breakfasts every other day (they have leftovers the other days). Oatmeal, rabbit feed (for a nice greens-based meal), raisins, scrambled eggs, cat food (protein—and they love it), apples, leftovers, mac 'n' cheese (a favorite!), green beans, and Cheerios (another favorite).

—Jennifer Kroll, White Lake, MI, Fluff 'N Strut Silkies owner

Chicken Feed Is a Personal Choice

The commercial feeds and university websites are geared for highest production per pound of feed, so naturally they're going to advise you to feed only a commercial-formulated diet. It's so easy to control exactly which nutrients go into the chickens using a commercial diet. However, if you watch chickens free range, they eat everything that can't outrun them.

—Barb Silcott, Milan, IN

Avian Tick Brigade

Let your chickens graze. My chickens keep our three acres almost totally free of ticks. Ticks for eggs—that's a really neat trade!

—Sharon Jones, Pine City, MN

Oats and Protein

In the summer, instead of feeding regular scratch, give your laying hens whole, plump oats. They like oats, they stay cooler, and they lay more eggs. Also, when my chickens need extra protein, like when they're molting or growing fast, I toss them a handful of dry cat food or meat-based dog kibble twice a day. They gobble it up and it adds lots of protein at a very low price.

—Marci Roberts, Springfield, MO

to a new location within the pasture. The free-range system is widely used throughout Europe for producing natural and organic meat chickens or eggs; it's becoming popular in North America, too.

Pastured poultry are raised in 8-by-10-foot portable pens with roofs but no floors and are confined day and night. Each pen contains about eighty chickens. Roosts and nest boxes are provided; feed and water are carried to the chickens. Structures are moved to fresh pasture (with the birds inside of them) once or twice every day.

Chicken tractors are engineered to move around your farm, usually by hand, to areas in need of enrichment. Size depends on the strength of the operator and ranges from $3 \times 5 \times 2$ feet for three or four chickens to a whopping $8 \times 12 \times 3$ feet. Usually, sides are crafted of wood-framed wire mesh; a hinged roof protects inhabitants from the elements and allows easy operator entry. If you build compact tractors to fit the width of spaces between your garden rows, the chickens will neatly weed and fertilize your garden without tanking up on produce while they do it (water and feed are provided inside the tractor; you'll need to advance the unit once or twice a day). Larger chicken tractors can be set atop spots needing more thorough cultivation over longer periods of time. Chickens are day laborers; they're removed from the tractor at night. It's a great way to cultivate and nourish garden soil while entertaining your chickens. There's a place for a chicken tractor (or two) on most every hobby farm!

This chicken tractor can hold eight to twelve chickens and allows them to choose whether they want to roost or to graze.

These free-range chickens are enjoying bug hunting in the sun.

chapter SEVEN

Chicks, with or without a Hen

Thee are few things more enchanting than a clutch of baby chicks. Chances are you'll want to grow some. There are three basic ways to start your chick collection. You can take the quickest and easiest route by buying "ready-made" chicks from a hatchery. If, however, you want to be involved from egg to newborn chick, you can choose the most labor-intensive route instead—incubation. Or you may decide in favor of the old-fashion approach to chick making—with a hen.

Whether you start with hatchery chicks or incubate your own (in an incubator or under a hen), you'll arrive at the same point: with a clutch of chicks to brood.

Hatchery Chicks

Most folks begin with hatchery chicks. They're inexpensive, readily available, and if you shop around, you'll find scores of breeds and varieties to choose from. See "Resources" at the end of this book for a list of mail-order hatcheries; reputable mail-order hatcheries are good sources for healthy chicks of the scarcer breeds. Don't assume you'll have to settle for Leghorns or Cornish Cross chicks when you shop close to home. Phone feed stores in your locale, especially prior to "chick time" (early spring), and ask what they'll be selling this year. Our old small-town feed store, Hirsch's in Thayer, Missouri, sells all the standard varieties, but it offers exciting rare breed chickies, too: German Spitzhaubens, Naked Necks, Black Sumatras, Dominiques, and Jersey Giants—even Red Jungle Fowl!

The difference between mail-order and local pick-up hatchery chicks is that the former are more likely to be stressed. However, handled correctly on arrival, healthy chicks from solid hatcheries rebound quickly. Most mail-order hatcheries tuck in an extra chick or two for each twenty-five you've ordered, to cover your losses in case a few don't make it.

Incubator Chicks

If you have access to fertile eggs and a sense of adventure, you can incubate your own chicks. You'll need an incubator and its accessories (a quality thermometer, a hygrometer to measure humidity, and a water pan if the incubator you choose doesn't have a built-in water reservoir). You'll need to build or buy an egg candling unit, too.

Good home incubators are expensive and incubating chicks is exacting work. If the atmospheric conditions inside the incubator are off for even a few hours, your eggs very likely won't hatch.

The Griggs family of Missouri uses this circulating-air cabinet incubator for chicken, guinea fowl, duck, and goose eggs.

Factoring in equipment costs, you can buy a lot of chicks for what it will cost you to hatch them at home.

However, when eggs in your batch begin pipping (when the peeping chickies peck holes in their shells)—what a priceless thrill! So if you want to try home incubation, here are the barebones basics.

Choosing and Maintaining Your Incubator

There are two types of incubators: forced air and still air. A forced-air incubator is fitted with one or more powerful internal fans that continually circulate air around the eggs. The ones used by hatcheries and commercial growers are monstrous things capable of incubating thousands of eggs set in stacked trays.

Tabletop models designed for hobbyists and small-flock poultry keepers hold from forty to one hundred eggs in a single tray and cost in the neighborhood of $75 to $575. Most are auto-turning units, meaning they turn the eggs for you at predetermined intervals. More expensive models come complete with one or more thermometers and sometimes a hygrometer.

It's easier to maintain constant heat and humidity levels—essential for good hatches—in forced-air units.

Temperatures are the same anywhere in a forced-air incubator; in still air models, temperatures stratify, so it's considerably warmer near the incubator's lid than it is on its egg rack or floor. Forced-air incubators are initially more expensive than comparable still units, but they're worth it.

A still-air incubator relies on vents set into its sides, top, and bottom for ventilation, which is not a particularly reliable system. Because ventilation is limited,

Here are close-ups of the Griggs family's circulating-air cabinet. Water pans inside the unit provide humidity essential for a proper hatch. However, not every egg hatches successfully. Bottom right, a chick pipped from this egg, but he couldn't break free of the shell's tough inner membrane.

it's more difficult to regulate heat and humidity in still-air incubators.

They cost from $12 to $16 for a tiny two-egg model (more toy than incubator, although with careful monitoring, it will certainly hatch eggs) to $85 for a top-of-the-line unit. A few still air models can be retrofitted with forced-air fans. It doesn't cost a fortune to get started with these units and, with care, they'll hatch a lot of eggs. However, it's very easy to overheat eggs and essentially cook them. It's a great deal harder to regulate heat and humidity in a still-air incubator.

Both types incorporate see-through lids or transparent observation ports, heating elements, built-in water receptacles or space to set a water pan, and egg racks. Automatic egg turners (which are very nice to have) are optional equipment on all but the most expensive models. Whichever you choose, don't toss the owner's manual! Each brand and model differs from the rest. Unless you follow the instructions in the manual provided with your incubator, you're unlikely to have much of a hatch.

It's important to set up the incubator indoors, in a draft free area, out of direct sunlight, and away from heat vents and window air conditioners. It must be thoroughly cleaned and disinfected before and after every hatch.

Two days before setting your fertile eggs, disassemble, scrub, and disinfect your incubator and accessories; then put it back together and fire it up. This gives it a chance to come up to heat and also gives you time to adjust the temperature and humidity.

Temperature, humidity, and ventilation must all be set properly and monitored to ensure successful development of the embryos.

Temperature

Place a trustworthy thermometer in the incubator at egg height. It should not be too close to the heating element and about 1 inch from the floor. In some units, the egg tray itself will do. Some folks play safe and set thermometers in two different locations.

If your incubator has a thermostat, set it at 99–100 degrees Fahrenheit for forced air units and 101–102 degrees for those without a fan. Inexpensive incubators usually lack thermostats; in that case, follow the owner's manual exactly—you must be able to reliably regulate the heat. A tip: if your incubator is heated with standard light bulbs, use bulbs of different wattages to adjust the heat; a 40-watt bulb is a good one to start with.

Don't add eggs until you've maintained the correct heat and humidity for at least eight hours. The acceptable range is from 97–103 degrees, but the closer you come to the exact degree, the better the hatch. Overheating is worse than the opposite; it's terribly easy to cook those eggs! Check often, at least twice a day. You'll find that your vigilance will pay off in chicks.

Humidity

To prevent moisture loss from your eggs—and to get them to hatch properly—you must control humidity scrupulously. Relative interior humidity should run 55–60 percent from days one through eighteen. For the last three days of incubation, humidity should be increased to 65–70 percent. Without a boost of additional moisture pipping chicks stick to their shells. Maintaining that higher humidity level is so vitally important that you must not open your incubator during the final three days of hatching, even to turn the eggs.

You'll need a hygrometer (also called a wet bulb thermometer) to measure the incubator's evaporative cooling effect. Buy one when you order your incubator or from poultry suppliers who sell incubators; a good one costs between $20 and $45. Some resources recommend wrapping cotton around the bulb of a regular thermometer to fashion a homemade version but it's "pound foolish" to improvise. Install the hygrometer so its wick (not its bulb) is suspended in water. To determine relative humidity, compare incubator thermometer temperature and hygrometer (wet bulb thermometer) readings.

HUMIDITY
Relative Humidity Calculator

Wet Bulb Reading						Incubator Temperature
97.7	89.0	87.3	85.3	83.3	81.3	100°F
91.0	90.0	88.2	86.2	84.2	82.2	101°F
92.7	91.0	89.0	87.0	85.0	83.0	102°F
70%	65%	60%	55%	50%	45%	= Percent Relative Humidity

This still-air incubator's windows allow observers to keep track of the action inside. The central switch adjusts the unit's temperature.

The greater the spread, the more evaporation is taking place. Ideally, hygrometer readings will run between eighty-three and eighty-seven. Raise the unit's relative humidity, especially just prior to the three-day hatching period, by increasing the evaporative water surface. Replace your existing water pan with a larger one, add a second pan, or prop one end of a soaked sponge in the existing water pan. You can also use an atomizer to spritz moisture through your incubator's vents.

Ventilation

Oxygen reaches the developing chick embryos via the fifteen thousand or so pores in the average eggshell; carbon dioxide exits the same way. Hatching eggs must breathe, or the chicks inside will die. As embryos mature, they require more fresh air. Incubator vents provide ventilation in various ways; read your unit's owner's manual to learn if, when, and how you should adjust the vents.

Manipulating the Eggs

A setting hen rolls her eggs—unintentionally, while shifting her weight, or intentionally with her beak—several dozen times a day. If she didn't do this, the embryos would stick to their shell membranes and die. With no hen in sight, it's up to you to do the egg turning (as the process is called) necessary for your incubator embryos to thrive.

It's also up to you to check that those embryos are indeed thriving by candling the eggs. Candling eggs is the process of holding an unhatched, incubating egg up to a strong light source to examine the contents inside and determine whether the egg is fertile. Infertile eggs and those that have quit growing need to be removed from the incubator before they begin to rot.

Egg Turning

Turn the eggs first thing in the morning, last thing at night, and at least once during the day; turning them more often is definitely better. Alternate the turns so the side that is down one night will be on top the following night. Use a soft lead pencil to mark one side of each egg with an "x" and the other with an "o" so you can keep track of the turning cycles.

Mark the eggs with an X on one side and an O on the other to help ensure even turning. A coding system is quite helpful since the larger the batch, the easier it is to lose track of your turning cycle. Even turning is imperative for a healthy hatch.

When placing eggs in still-air incubators, position them on their sides with their small ends tipped slightly downward; rotate them one-half turn at least three times a day. In forced-air incubators, place them small end down and tilted in one direction; tilt them the opposite direction to turn them.

Do not turn eggs during the last three days before hatching. Chicks will be maneuvering into pipping position, and moving disorients them. Besides, to maintain constant 65–75 percent relative humidity throughout the pre-hatch period, you don't want to open the incubator for any reason.

Some incubators incorporate automatic egg turners and many that don't can be retrofitted with the feature. This is must-have equipment, especially if you can't be there to turn the eggs yourself.

Candling Eggs

Candling doesn't harm the growing embryos as long as you work quickly.

Most folks candle eggs at seven days and again one week later. You'll need a dark room and a candling device. These can cost less than $20 (for a basic hand-held model) to $350 or more. But for incubator newbies, a homemade version works nicely indeed. Turn out the lights, switch on your candler, and then one by one, hold the eggs in front of the intense light.

Infertile eggs appear empty, with only a shadow of the yolk suspended inside. You'll spy a spider-like clump of dark tissue in fertile eggs: a glob with blood vessels radiating out from it like spokes. "Dead germs" are eggs that were fertile but have died; in such eggs, a faint blood ring circles the embryo. If you're not certain you're seeing one, return the egg to the incubator; you can catch it next time.

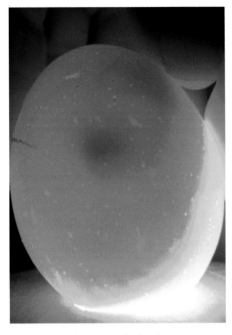

Here is an egg being candled at day seven. Note the embryo's U-shaped shadow and dark eyeball.

White eggs are easier to work with than brown or speckled ones, and the second candling is trickier than the first. Is that embryo alive or not? It can be hard for a newbie to tell. Check for candling topics at the websites listed in the Resources section.

Incubating Timetable: Preparation to Pipping

First you should buy or collect hatching eggs. Two days prior to setting them, scrub the incubator squeaky clean. Use a weak bleach solution or a commercial product to disinfect the unit and all supplies. Haul the incubator to the spot where it will be parked for the twenty-one-day hatch cycle, then fire it up.

Spend the next two days (or more) fiddling with heat and humidity levels until everything is perfect.

Six hours before you plan to set them, remove your hatching eggs from storage (see page 93) and allow them to come slowly to room temperature. Once incubation commences, religiously monitor conditions inside the unit. Make minor adjustments as needed. Turn the eggs at least three times a day, but don't open the incubator any longer or more often than you absolutely have to. Candle the eggs on days seven and fourteen. Discard the ones you're reasonably certain won't hatch.

Power Outages

If your power goes out, don't abandon the hatch. Carry your incubator to the warmest place in your house. If it's a forced-air model, crack the lid to admit fresh air; it's better to keep still-air models closed.

Should the power come on within a few hours, resume incubation, but candle the clutch four to six days later and discard any eggs that appear to have died. Surviving eggs require more time to hatch, sometimes as much as one or two days.

If 50 percent of the eggs you set hatch, celebrate! A lot can go wrong during the home incubation process; a 50/50 hatch is a good one, indeed.

Incubation Troubleshooting

Problem	Possible Cause
No embryonic development (infertiles)	1) Not enough roosters 2) Aged rooster(s) 3) Temporarily infertile rooster(s) due to frozen comb and wattles 4) Eggs stored too long or incorrectly
Blood rings ("dead germs"); Early embryonic death	1) Incorrect or fluctuating temperatures 2) Eggs stored too long or incorrectly
Many dead early-stage embryos	1) Incorrect or fluctuating temperatures (usually too high) 2) Insufficient ventilation 3) Improper egg turning
Fully formed chicks/embryos died before pipping	1) Incorrect temperature 2) Insufficient humidity 3) Insufficient ventilation 4) Improper egg turning 5) Eggs chilled prior to incubation
Fully formed chicks/embryos died during pipping	1) Insufficient humidity 2) Insufficient ventilation 3) Improperly positioned eggs resulting in malpositioned chicks
Early hatch	High temperature
Late or uneven hatch	1) Low temperature 2) Uneven temperature in incubator 3) Old or improperly stored eggs
Chicks stuck to shells	1) Humidity too high during early incubation 2) Humidity too low during late incubation (especially during the three days of hatch)
Crippled chicks	1) Incorrect temperature (usually too high) 2) Low humidity 3) Improper egg turning 4) Insufficient traction on hatching tray
Small, abnormal, or weak chicks	1) Small eggs hatch small chicks 2) High temperature or low humidity 3) Insufficient ventilation 4) Diseased breeder flock 5) Poorly nourished breeder flock
Large, mushy, weak chicks	1) Low temperature 2) Insufficient ventilation
Rough or unhealed navels	High or fluctuating temperatures

What to Do

1) Add more roosters
2) Add younger roosters
3) Provide warmer lodgings in midwinter; add fertile roosters to the flock
4) Correct improper storage methods

1) Monitor temperature readings more closely
2) Correct improper storage methods

1) Monitor temperature readings more closely
2) Increase ventilation/air circulation
3) Turn at least three times a day

1) Monitor temperature readings more closely
2) Monitor hygrometer readings more closely
3) Increase ventilation/air circulation
4) Turn at least three times a day
5) Gather eggs often; correct improper storage methods

1) Monitor hygrometer readings more closely
2) Boost ventilation/air circulation
3) Set eggs small-end down. Don't turn them during the last three days of incubation

Operate at recommended temperature; check accuracy of thermometer(s); move incubator to a cooler room

1) Operate at recommended temperature; check thermometer's accuracy
2) Contact incubator company for repair or replacement
3) Correct improper storage methods

For both: monitor hygrometer readings more closely

1) Monitor thermometer readings more closely
2) Monitor hygrometer readings more closely
3) Turn at least three times a day
4) Use wire floor trays or cover existing tray with cheesecloth

1) Discard small hatching eggs
2) Monitor thermometer and hygrometer readings more closely
3) Increase ventilation as needed
4) Use eggs from disease-free flocks
5) Boost flock nutrition

1) Monitor thermometer readings more closely
2) Increase ventilation/air circulation

Monitor temperature readings more closely

Three days prior to the expected hatch (on the eighteenth day of incubation), turn the eggs for the final time, and then crank up the heat and humidity. Don't open the incubator again until the hatch is complete. One day before the hatch, clean and disinfect pre-used chick waterers and feeders. Set up your brooder and switch on the heat.

When pipping begins, don't help chicks out of their shells. Opening the incubator compromises the rest of the hatch and a chick who can't break free is nearly always crippled or too weak to survive. Make sure to leave the chicks in the incubator for at least twenty-four hours, certainly until they're fluffy and dry. Taking them out early subjects them to chills. After twenty-four to thirty-six hours, remove the chicks to a warm box and carry them to their brooder.

Hold remaining eggs to your ear and listen closely. If they're going to hatch, you'll hear movement and possibly cheeping; put those eggs back in the incubator and discard the rest of the unhatched eggs. When the hatch is complete, disassemble, scrub, and thoroughly disinfect the incubator and its accessories before packing them away.

Chicks the Old-Fashioned Way

A cheaper and easier way—and often a more successful one—is to hatch eggs under a broody hen. Not all hens brood eggs, and none of them do it all the time.

Because a setting hen lays no eggs and won't lay again for awhile after she hatches chicks, broodiness has been bred out. For example, Leghorns and Leghorn crossbred super-layers almost never brood. At the other end of the spectrum, some Silkie and Cochin bantam hens set and hatch chicks at the drop of a hat. Most bantam, dual-purpose, and heavy breed hens (especially Asiatics such as Brahmas and Cochins) will set; Mediterranean and Continental breeds don't often tend to brood.

The Hen

If a hen lingers in a nest box after laying, if she ruffles her feathers and sputters when you take her egg, lift her off the nest and set her on the floor. Check on her later. If she's on the throne again, you've probably got a broody in the making.

She can set on the eggs she's accumulated if you allow it, or on eggs of your choosing, but you shouldn't let

Broodiness by Breed*

Broody

Araucana, Ameraucana, Australorp, Belgian d'Uccle, Brahma, Chanteclar, Cochin, Cubalaya, Dominique, Java, Langshan, Old English Game, Orpington, Silkie, Sumatra, Sussex, Wyandotte

Largely Non-Broody

All hybrid layers and Ancona, Andalusian, Campine, Hamburg, Houdan, Leghorn, Minorca, Spanish White Face

*Not every individual of any breed behaves absolutely true to form. The rarest of Silkies will run screaming from nest eggs, while a few hybrid layers will set.

My husband, John, just removed Gracie from her nest and she's furious. His thick gloves are more than a precautionary measure; an enraged broody hen will surely peck at the hands invading her space.

her set them in the henhouse because other chickens will pick on her. And if she leaves her nest for her daily constitutional and returns to find an interloper on her eggs, there will be a brawl—eggs can be shattered and the broody might lose. It's best to set her up in her own private lodging.

Hens prefer cozy, secluded cubbyholes in which to set. Build your broody a disposable getaway by slicing ventilation slots that are 1 inch from the tops of all four sides of a lidded cardboard box that is slightly larger than the hen. Generously pad the bottom with chopped wheat straw or shavings.

Hollow out a bowl-shaped depression in the nesting material, and then wait until dark to relocate your hen. When you do, wear gloves—she'll peck, *hard*!

Reach under the hen and remove her eggs. Nestle them into the broody box nesting material, then move the hen. Slide your hand beneath her, fingers facing up; wait until she quits fussing and settles down onto your palm. Then transfer her to her temporary home and close the lid.

Place the box in a predator-proof, dimly lit spot. Get a small, deep container of water and cuddle it into the litter in one corner of the box; place a shallow bowl of feed in another (soup and tuna cans work well). Don't open the lid again until late the following day. This will give your hen time to acclimate to her new home, and she'll appreciate the solitude during this period. She will have food and drink if she wants it (she probably won't, but giving her the option will make you feel better), and she won't mess her nest just because she can't get out. If she seems content, cut an entryway in one side of her haven, then pat your back: you've successfully resettled a hen!

Once she begins setting, a hen's physiology changes. Her temperature drops to around 100 degrees, and her metabolism slows. She molts feathers from her breast and underpinnings, the better for her skin to warm eggs and, later, chicks (the molted feathers would act as insulation if she didn't lose them). She remains on her eggs, feathers ruffled and wings slightly extended to warm them, full-time until they hatch, save a half hour or so every day or two to drink, grab a bite to eat, and make 'broody poop' (a memorably stinky, single, gigantic glob of birdy-doo).

She'll set until her chickies hatch or until it's obvious that they won't. If she's still setting on unhatched eggs after twenty-three days, day-old hatchery chicks or another hen's newly-hatched babies can be fostered under her. Salt her nest with chickies at night; by morning, chances are she'll decide they're her very own.

The New Chicks

Don't allow a hen with brand new chicks to free range or to immediately rejoin the main flock. Other chickens may harry her or the babies, and when the little ones trail mama through wet grass, they easily chill. Chilled peeps are likely to die. Cats, hawks, skunks, and their kind adore chickie aperitifs, so if you want them to live, house those chicks and their mama in their own separate coop.

Babies should be fed a chick-starter ration inside of a creep feeder. This is a structure that only they can enter, because its openings are too tiny for mama to squeeze through. An over-turned heavyweight cardboard box with chick-size openings carved into its sides makes a fine, free creep feeder. Weight the box top with a brick or flat stone just heavy enough to keep mama from tipping it over.

This little guy is the first of his clutch to pip, and the neighboring eggs aren't far behind. Avoid the urge to help a pipping chick out of his shell. If he lacks the strength to emerge, he won't be suited for life on the farm.

Make certain the peeps can always reach fresh water. Place a chick waterer in the brooder pen. Drop some marbles into the drinking surface to prevent the chicks from drowning, and keep it clean.

When they're a month old and nicely started, the chicks and their mama can join the flock. But monitor things for a while; some adult chickens are mean to other hens' chicks.

Hatching Eggs 101

Whether you hatch chicks under a real, live mama or inside a "tin hen," you'll want to begin with quality eggs. Following are several requirements for quality eggs.

Fertile. Hens don't need a rooster's input to lay eggs, but if you want fertile eggs, you definitely need the services of Mr. Roo. He shouldn't be too young—or too old; he also should be actively breeding hens. You need one prime rooster for every eighteen bantam, twelve light-breed, or eight heavy hens in your flock.

Clean. A poop-smeared egg can spread fecal-borne disease to the embryo within and to other, cleaner eggs in the clutch. Scrubbing won't help. Bathing removes the protective natural sealant present on newly-laid eggs and is apt to force bacteria through the egg's porous shell. Some folks lightly sand away splots with finest-grain sandpaper, while others believe that doing so weakens the shell. Is it worth the risk? Probably not, especially if other, cleaner eggs are available.

Average size. Huge eggs don't hatch well; undersized eggs hatch undersized chicks. An average size is best.

"Egg shaped" and intact. Toss cracked eggs, misshapen eggs, and any with shells that are wrinkled, rough-textured, or thinner than usual.

Promptly gathered. Collect hatching eggs first-thing every morning and recheck nests throughout the day. Don't let them get chilled, overheated, or unnecessarily soiled due to being in the nest for too long.

Properly stored. Fertile hatching eggs should be placed in egg cartons, small end down, and tilted to one side, then stored in a cool (50–60 degrees Fahrenheit), humid (70–75 percent) place in your house. Never keep hatching eggs in the refrigerator! A simple way to turn them (you only need to do this once a day) is to slip a 1-inch slice of a two-by-four under one end of the carton, then move it to the opposite end the next day. Keep it up until you're ready to set the eggs. Don't store them longer than ten days—and less time is a whole lot better.

When You Don't Want Chicks

Not everyone wants their hens to brood. Setting hens, hens raising their chicks, and post-broody hens who didn't hatch a brood (because their eggs were infertile, a predator snatched them, or for any other reason) don't lay eggs. And the longer she broods, the longer the interval before she'll lay again.

Furthermore, not every hen makes for the very best mom. Some abandon their nests mid-incubation, while others are so scattered that they stomp on and shatter their eggs. Less maternal souls sometimes hatch their brood successfully but dislike their chicks; they'll peck their chicks' little noggins—and worse—some kill their babies (and eat them!).

Free-spirit hens gallivant off and leave their babes to their own devices. Whatever causes these hens to behave this way, they shouldn't be allowed to set eggs, unless you're willing to snatch their newly hatched chicks and raise the peeps in a brooder.

If you want to turn your broody hen's mind to laying instead of hatching chicks, you have to "break her up." Breaking up a broody sounds easy—just cart her eggs away and off she goes—but it's not; most broodies continue setting, eggs or not. And sometimes it goes on for a good, long (nonproductive) time.

If you've removed her eggs but your hen stoutly insists on setting, try some proven ploys to get her to stop. First, after nightfall, pull on your gloves, hoist Ms. Broody up out of her nest, and resettle her in a completely new location for three or four days (or more, if needed). Disassemble the nest before you release her. If it's disposable, toss it. If she was holed up in one of your laying nests and you moved it for her, scrub and disinfect it, refill it with fresh nesting material, and take it back to the henhouse where it belongs.

If that doesn't work, try temporarily housing her in a "broody coop," a wire mesh cage (a small all-wire rabbit hutch works well) suspended from the roof or a rafter, where cool air circulating around

These babies stay close by their mama's side. Wait at least a month before allowing a hen and her chicks to join the flock to avoid possible scuffles between veterans and youngsters.

the hen's torrid underpinnings is likely to break through her single-minded trance. One caveat: make certain you don't hang her in a draft! Three or four days in the broody coop should work its magic. Placing two handfuls of ice cubes under her several times a day or bathing her in cold (but not icy) water, then carting her off to short-term incarceration, should also speed along the process and cool her underpinnings, too.

Whatever you try, persevere: you can outlast a hen! Provide her with plenty of food and water, despite what a few old-timers may tell you. Most seasoned chicken keepers claim starving broodies makes them surrender sooner, but starving her will weaken an already stressed bird, and she may even die.

To discourage hens from going broody in the first place, choose a breed that doesn't tend to set. Promptly remove fresh eggs from nest boxes; encountering an inviting collection of eggs triggers the urge to brood in many good hens of the broodier breeds.

Brooding Peeps

So now you have chicks. Whether from a hatchery, from your incubator, or out from under a hen, they're yours to raise. It can be tricky, but if you stick to the rules, brooding peeps isn't hard. To have fun doing it and to lose fewer chicks, here's what you have to know.

The Brooder

If there is a brooder house on your farm, fine and dandy. If it hasn't been cleaned since its last occupation, strip the floors and disinfect them. Sweep down cobwebs and disinfect the lower walls, too.

Decide where to hang your heat lamp (more about this later) and erect a cardboard draft-protection shield beneath it. Use a commercial product (mail-order

These young birds from Gib and Melba Mullins' Missouri farm aren't quite old enough to leave their brooder box to join the rest of the flock. A brooder provides a warm environment for newly hatched chicks to grow during their first month.

Tabletop Brooder

To make your own tabletop brooder, you will need:

- A hard plastic storage box, preferably with translucent sides. You'll find a huge variety to choose from at Target and Wal-Mart. Figure necessary floor space per chick before you go shopping. If one box isn't enough to house them through the brooding period, start out with all the chicks in one, then transfer part to another box as the brood matures. You can count on the following space requirements per chick: through the end of week three, 6 square inches; weeks four through five, 9 square inches; and weeks six through eight, 1 square foot.

- A sharp knife (an Exacto knife works well)

- ¼ inch mesh hardware cloth—a piece slightly smaller than the box's lid

- 10 bolts, ¼ by ½ inch in length; washers and nuts for each (this may vary depending on lid design)

- A sharp nail or awl to use as a scribe

- Wire snips

- A drill

- A screwdriver

First, invert the box top on a hard surface. Scribe a 1-inch border inside any framed panels (some lids feature a single panel, others have two). Using the sharp knife, cut a window or windows into the lid. Allowing a 1-inch overlap, snip the hardware cloth screen to fit, drill holes to secure it, and bolt it down. That's it!

Place a gooseneck reading lamp on the screen to use as a brooder lamp. Fiddle with standard lightbulbs of various wattages until the interior temperature is perfect. (See "Broiler Requirements on page 99.) Fine-tune temperatures by raising or lowering the lamp. Switch to hotter or cooler light bulbs if needed. Nothing could be easier!

Use our easy plans in "Tabletop Brooder" to build this translucent plastic-box brooder.

hatcheries and poultry supply houses sell them) or a homemade version fashioned from panels that are 12–18 inches high and snipped from cardboard boxes and connected with strips of duct tape. A 6-foot circle is fine for fifty chicks; if you have fewer than that, it's easier to brood in a box and transfer chicks to the floor later on.

Most small-scale chicken raisers don't own brooder houses. Initially, it doesn't matter. Today's favorite chick brooder is a secondhand cardboard box stowed in your home in a warm, out-of-the-way spot. A semi-topless cardboard box (leave the flaps on so you can regulate ventilation) is clean, dry, and draft-free. When the chicks outgrow it or it gets smelly, simply dispose of it. Transfer the chickies to a new box and shred the old one for the compost pile.

Other ingenious homemade brooders can be fashioned of flexible plastic wading pools, old aquariums, dog crates, or rabbit hutches fitted with cardboard draft shields. My personal favorite: inexpensive plastic storage-box brooders like this one.

Or you can always buy a ready-made brooder. A traditional galvanized-steel box-type model with mesh floor, built-in water and feed troughs, and its own heating unit costs in the neighborhood of $200 and houses fifty chicks for up to fourteen days.

Stromberg's plastic brooder, which has a twelve-chick capacity and is made of high-impact plastic, is a better bet for pet and small flock owners. It resembles an airline pet carrier, breaks down easily for cleaning, and heats nicely with a 75- to 150-watt incandescent lightbulb. At less than $35 postpaid, it's a steal!

The Furnishings

Whichever type of brooder you have, you'll need to furnish it with a heat source, litter, feeders, and waterers. Consider the following when choosing which kinds to buy.

A utility lamp provides essential heat for these newborns. Store-bought brooder boxes may come with a heat source, but homemade versions will require a bit of creativity. Be sure your choice is securely fastened and is of the correct wattage to prevent injury or death.

Heat Sources

Ready-made brooders usually incorporate their own heaters. If yours doesn't, or if your brooder is the homemade kind, you'll need a reliable heat lamp to warm those tender chickies or they won't survive.

Infrared heat lamps, which come in clear- and red-bulb versions, provide the constant heat that wee chicks require. Red bulbs throw less light and are said to prevent juvenile picking that sometimes leads to cannibalism (we'll discuss this charming habit in a bit). When choosing an infrared lamp, opt for a UL-approved model with a porcelain socket and a lamp guard. If reusing an older unit, make certain its cord hasn't frayed.

Used improperly, these lamps can burn down barns, brooder houses, and homes, so make absolutely certain the lamp can't fall or overheat nearby flammable surfaces. Hang it by a chain, not by its cord! You'll need one lamp fitted with a 75- to 150-watt bulb per fifty to seventy-five peeps.

Storage-box and aquarium brooders can be heated using everyday lightbulbs. You'll need a variety of wattages in the 75-watt range in order to adjust the temperature. A gooseneck table light makes a handy fixture for tabletop brooders. Whichever type your heating unit requires, don't forget to buy extra bulbs!

Litter

The litter you choose makes a world of difference. It must be insulative, absorbent, and provide lots of grip. Lodging new chicks on slick sheets of newsprint or flattened cardboard makes their tiny legs slip to the sides, causing *spraddle leg* (a disability in newborn chicks caused by an inability to properly grip a surface with their feet). Spraddle-legged chicks can sometimes be salvaged by hobbling them with Band-Aids or makeup sponges until their legs correct, but spraddle leg is a problem better prevented than cured.

Sawdust sometimes confuses new chicks, who think it's food. Ingested sawdust leads to *pasty butt*—droppings stuck to a tiny chickie's tush. If you don't soak or pick it off, the chick can't eliminate and it'll quickly die. Sawdust litter should not be used until chicks are a few weeks old.

First-class litter materials include pine shavings, coarsely ground corncobs or peanut shells, rice hulls, peat moss, sand, and old bath towels weighted down at the corners and laundered whenever they get soiled. Don't use hardwood shavings; some species are toxic to tiny chicks.

Initially bed the brooder area with 3–6 inches of litter, more if it's chilly outdoors and the brooder sits directly on a floor. Stir and fluff litter every day. Scoop water spills and messes as they occur. Add bedding whenever needed to keep things cozy and tidy.

Feeders

Flying saucer-shaped galvanized steel feeders with hole-studded snap-top lids are ideal for tabletop brooders. Folks with more chicks to brood will probably prefer trough feeders (allowing 2 feet of feeder per each of twenty-four chicks).

Empty pressed-fiber egg cartons make feeders for tiny chicks that are both easily accessible and disposable, although they'll climb on and poop in them, wasting a lot of feed in the process.

If you're raising meat chicks, choose 20–22 percent protein broiler starter for them; if they are pets or future layers, standard 18–20 percent protein chick starter is a wiser choice.

Most starters are medicated with Amprolium to prevent coccidiosis until the chicks develop their own immunity to this common disease. Some feed is

Brooder Requirements

Age	Temperature	Floor Space Per Chick	Waterer Space Per Chick	Feeder Space Per Chick
Week 1	95 °F	3 sq. in.	0.5 in.	1 in.
Weeks 2–3	85–90°F	6 sq. in.	0.5 in.	1.5 in.
Weeks 4–5	75–80°F	9 sq. in.	0.75 in.	2 in.
Weeks 6–8	70°F (Higher if chicks seem chilly)	1 sq. ft.	0.75 in.	2 in.

While this mason jar can hold a lot of feed, we only fill it halfway at most. Chickens bill out excess feed and spoil it by soiling in it and walking on it. Over time, their wastefulness adds up to a lot of unnecessary expense.

fifty chicks. Chicks easily drown in waterers, so whichever type you choose, add marbles or pebbles until only their beaks can get wet. When you place units in the brooder, don't position any near the heat source; chicks don't fancy warm water and might not drink at all if it's hot. Empty and brush-scrub waterers every day; rinse them with a weak bleach solution once a week. Make certain waterers are filled and functional at all times.

Many veteran chicken raisers spike peeps' drinking water with table sugar (one-quarter cup sugar per gallon of water) to give them a needed energy boost. Others swear by vitamin and electrolyte supplements such as Murray McMurray Hatchery's Quik-Chik and Broiler Boosters. If you plan to use water supplements, lace the peeps' drinking water right from the start.

The rocks in this waterer prevent tiny chicks from drowning. This model contains electrolyte- and vitamin-laced water, which gives baby birds a jump-start on proper nutrition.

laced with antibiotics, too. Non-medicated feed is available, but you'll have to ask for it—and likely pay a premium price—if you want it.

Unless you supplement your chicks' diet with scratch, they won't require grit. If you do choose to supplement the food, buy special chick grit or cage-bird grit (available from a pet store).

Waterers

If your brood is a small one, you'll need one plastic-based quart-size canning jar waterer for each dozen chicks. For larger broods, one gallon-size waterer services

Can't tell chicks apart? We mark ours with a non-toxic ink pen or food coloring. Don't use red—chicks instinctively pick at it.

Putting It All Together

At least twenty-four hours before anticipated hatch or delivery, set up your brooder, switch on the heat lamp, and bring everything up to heat. The temperature at chick height (2–3 inches from the floor) must run a constant 95 degrees Fahrenheit for the first full week; use a thermometer to check it twice a day. You'll lower the temperature by about 5 degrees each week until the chicks are five weeks old. After that, maintaining heat at 70 degrees (or indoor room temperature) generally does the trick.

Chicks instinctively pick at whatever they see, so for the first few days, carpet their litter with nubby paper towels. Sprinkle a thin layer of chick starter on it to encourage the chicks to pick at feed instead of litter. Tempt slow-learning chicks with chopped boiled egg yolk sprinkled on paper towels or drizzled atop their regular feed. Egg yolk is the ideal chickie appetizer!

Listen and Look

Listen to your chicks. Contented chicks converse in gentle cheeps; frantic, shrill peeping means they're chilly. Up the heat by moving the lamp closer to the chicks or substitute a higher wattage bulb. Keep an eye on your chicks, too.

Happy chicks spread throughout the brooder, allowing each other plenty of space. Chilled chicks cluster beneath the heat source, sometimes piling atop one another and suffocating the babes at the bottom of the heap. Overheated chicks extend their teensy wings and pant; they flock to the outer edges of the brooder to flee excessive heat. When the peeps huddle at one side of the heat lamp or another, suspect a draft.

Ongoing Care

Once chicks are eating well, fill feeders halfway full (to reduce waste), but never let them run out of water or feed. If feed gets damp or dirty, dump it; rinse and dry the feeder before replenishing it with fresh starter.

Carefully and frequently handle future pets, but only for a few minutes at a time; otherwise it's best to leave tiny chicks in the brooder. Supervise children; remind them that the peeps are fragile, and make sure kids wash their hands after handling chicks.

Nip toe- and feather-picking in the bud. Chicks pick each other when they're too hot or too crowded, when their light is too bright, if the air is too stale or their feed inadequate, or sometimes simply because they feel like it. Picking leads to cannibalism—not a pretty sight.

Chicks instinctively peck at everything in their environment, so if yours begin picking one another, add grass clippings to their diet. Strew bits of healthful greenery around the floor and let them pick at that, giving the chickies something constructive to do. Switch the clear heat lamp or lightbulb for a red one (which should have a calming effect).

Remove picked chicks to safer quarters and dab their wounds with anti-pick solution, such as Hot Pick, Blue Kote, or an old favorite, pine tar, to heal them and deter further picking.

As chicks mature, provide additional floor space, feeders, and waterers. If brooding in a brooder house, remove the draft shield when the chicks are three weeks old. By the time they're six weeks old, chicks are fully feathered and fit to face the world. It's time to move pullets to the henhouse and meat chickens to quarters of their own.

Chicks will pick at anything; it's better that they pick at a variety of food than at each other.

Tiny Incubator Can Due!

I hatched our Li'l Due in a $15 incubator. I started with three eggs, but he was the only one who hatched. I had to help him from the shell, and he had splayed legs, but I put an adhesive bandage on them, and he only had to wear it one day until he was fine. He's almost ten months old now. The little incubators do work.

I had to keep the room at a steady temperature, though. If the room got warmer, so did the incubator. All I had was the little thermometer that came with it and the little plastic wrap reflectors. I didn't have a water weasel; I just kept the feet filled with water in the incubator. It may have been beginner's luck, and there was certainly a lot of prayer involved, too!
—Patty Mousty, New Albany, IN

No Incubator?

I think it would be better for new chicken people to buy chicks. There are many things that can go wrong when you incubate and that could discourage a new chicken owner. They have spent a lot of money and will never use the incubator again.

Chicks are cheap, you can get them sexed or as sex-links, and you are going to get the right number or close to the right number. Brooder box management is hard enough to start with. Start with the cute fuzzies, then do eggs. After you have chickens awhile,

then you can decide if you want to incubate.
—Helen Jenson, Silverton, OR

Treasure That Hen

When you find a hen who likes to hatch eggs and does a good job of it, guard her with your life! You can set Guinea eggs, ducks eggs, even goose eggs under her if she's big enough. Ducks or geese with a chicken mom are such fun. It drives the chicken mom crazy when her chicks make a beeline for mud puddles!
—Marci Roberts, Springfield, MO

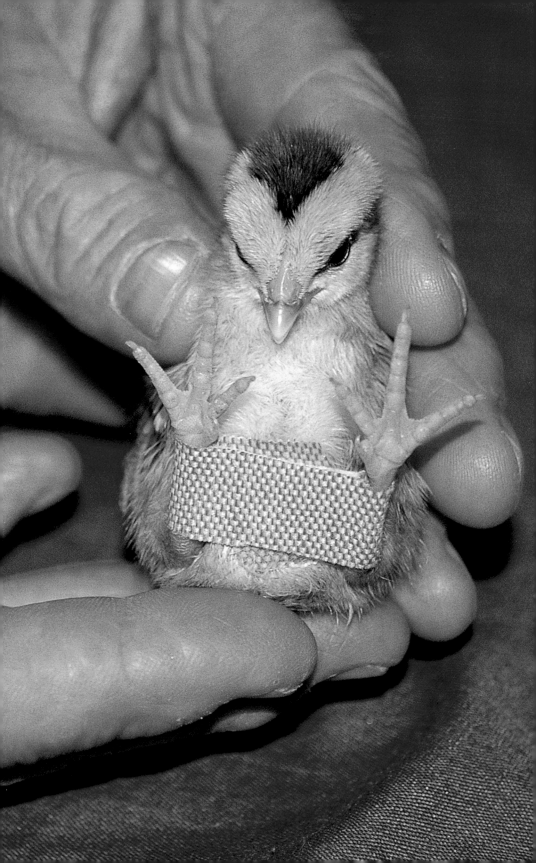

chapter **EIGHT**

Chickens
as Patients

Most introductory chicken manuals fail in their attempt to describe the thousand and one diseases, maladies, and afflictions that sometimes befall our friend, the chicken. This one isn't even going to try. The amount of information we could pack between these pages would be scant indeed, compared with the plethora of readily accessible, specialty material available on the Internet. Check out the resources in the back of this book, print what interests you, and stuff your printouts in a binder: create your own poultry veterinary book.

Preventing Problems

It's infinitely easier to keep chickens healthy than to doctor them after the fact, and in most cases, keeping them healthy isn't at all hard. To keep healthy, happy, bright-eyed chickens, begin with healthy, happy, bright-eyed chickens.

Buy from reliable sources; don't stock your coop with someone else's rejects or bedraggled bargain fowl picked up at country swap meets. Consider starting with day-old chicks from reputable hatcheries and know what you're working with right from the get-go. Maintain a closed flock. Don't indiscriminately add chickens to your collection. It upsets the flock's hierarchy and causes infighting and stress. It's also the best possible way to introduce disease.

Provide suitable quarters. They needn't be fancy, but they must be clean, roomy, well-ventilated, and draft free. Feed your chickens what they need to thrive. Keep plenty of sparkling clean drinking water available. Furnish enough feeders and waterers to make certain every chicken can eat or drink whenever it pleases.

Avoid unnecessary stresses. For example, keep your kids or dog from chasing the chickens, handle them gently, make changes gradually, and don't upset the status quo. Laid-back chickens tend to be healthy chickens.

It's important to recognize problems early, while they're still fairly simple to fix. Consult the "Is My Chicken Healthy" chart for general signs of health and ill health.

This Black Jersey Giant has twisted toes, a defect with both dietary and genetic roots.

Maladies: Parasites and Diseases

A parasite-savvy chicken raiser knows early detection is a key to keeping chickens healthy. Most chickens are exposed to wild birds, which commonly spread parasites and disease. Left unchecked, parasites can spread like wildfire through your flock, causing anemia, weight loss, decreased egg production, and even death.

Parasites, Lice, and Mites

Chickens sometimes do have worms. Zap external and internal parasites before they cause problems. However, don't rely on over-the-counter remedies that treat parasites your chickens might not have. Instead, collect a community fecal sample using material from a number of chicken plops and take it to your vet to be checked for worm eggs. If worms are present, she'll prescribe a wormer custom tailored for your flock. Do this twice a year to keep your birds in the pink.

Lice and mites are the scourge of the henhouse, and you need to be vigilant and take care of them. The first step is to know what you're dealing with. See "Chicken Itchies" in this chapter for a description of the big four.

For the scoop on external creepies, scope out the University of Florida's bulletin, "External Parasites of Poultry" at http://edis.ifas.ufl.edu/ig140. Before applying any commercial bug blaster, read the label carefully and follow instructions exactly.

Communicable Poultry Diseases: The Big, Scary Ones

It's unlikely your pet chicken or small farm flock will succumb to one of the big, exotic poultry diseases currently in the news. However, these diseases are out there, so you need to be able to recognize their symptoms. If you do suspect one such disease has a toehold in your flock, bring in the big guns. Most of these are diseases that must be reported to health authorities, and it's your duty to wave the red flag. (See the "Major Chicken Maladies" chart in this chapter for a summary of the diseases listed below.)

Avian Influenza: Avian influenza is a highly contagious respiratory infection caused by type A influenza orthomyxoviruses. Symptoms vary widely and range from mild to serious. Death occurs suddenly; the disease sweeps through a flock in just one to three days. Avian influenza occurs in many forms worldwide and must be reported in the United States.

Fowl Cholera: Fowl cholera is an acute, relatively common disease caused by *Pasteurella multocida* bacteria. It spreads rapidly and kills quickly; chickens die within hours after symptoms appear. Humans handling infected birds sometimes contract upper respiratory infections.

Infectious Coryza: Infectious coryza is found worldwide; in the United States, most cases occur in the southeastern

Is My Chicken Healthy?

What to Check	Healthy	Unhealthy
Temperature	103–103.5°F	Higher/lower
Respiration	Easy, even	Labored, rasping, coughing
Posture	Stands erect, head and tail elevated; alert and active	Hunkered down, tail and wings drooping; depressed
Appetite	Eats often	Uninterested in food
Thirst	Drinks often	Excessive thirst
Manure	Formed mass, gray to brown with white caps; fecal droppings may be frothy	Liquid or sticky; green, yellow, white, red
Condition	Feels heavy, firm, powerful	Thin, weak, thin-breasted
Feathers	Smooth, neat, clean	Ruffled or broken, dirty, stained
Comb and Wattles	Bright red; firm	Shrunken, pale, or blue
Eyes	Bright and alert	Dull, watery, possibly partially closed
Nostrils	Clean	Crusty, caked
Legs and Feet	Plump, scales clean and waxy smooth, warm joints	Enlarged, crusty scales; hot swollen joints; soles of feet crusty, cracked, or discolored

Chicken Itchies

Poultry Lice

While there are many forms of poultry lice, the most common are the chicken body louse (*Menacanthus stramineus*) and the shaft louse (*Manopan gallinae*).

- Fast-moving, six-legged, flat-bodied insect with broad, round head; 2–3 mm long, straw-colored (light brown)
- Female lays 50 to 300 tiny white eggs near base of feather shafts
- Does not suck blood; feeds on dry skin scales, feathers, scabs; gives infested birds a moth-eaten look
- Spends entire life cycle on host
- Primary infestation seasons are fall and winter
- Dust or spray birds and their environment using commercial products such as Malathion, Permethrin, Rabon, or Sevin, carefully following the instructions on the label. Always consult with your veterinarian before using such products.

Chicken Mite (also called Red Roost Mite)
Dermanyssus gallinae

- Slow-moving, eight-legged insect; 1 mm long (the size of coarsely ground pepper); gray to dark reddish brown
- Lays white or off-white eggs on fluff feathers and along larger feather shafts
- Sucks blood only at night; hides in cracks and crevices in coop or poultry building during the day
- Primary infestation season is summer
- Dust or spray birds and their environment using commercial products such as Malathion, Permethrin, Rabon, or Sevin, carefully following the instructions on the label. Always consult with your veterinarian before using such products.

Northern Fowl Mite
Ornithonyssus sylviarum

- Slow-moving, eight-legged insect; 1 mm long (the size of coarsely ground pepper); brown
- Lays white or off-white eggs on fluff feathers and along larger feather shafts located on host's vent, tail, back, and neck
- Sucks blood
- Spends entire life cycle on host; feeds day or night
- Primary infestation seasons are fall, winter, and spring

- Dust or spray birds and their environment using commercial products such as Malathion, Permethrin, Rabon, or Sevin, carefully following the instructions on the label. Always consult with your veterinarian before using such products.

Scaly Leg Mite
Knemidokoptes mutans
- Slow-moving, eight-legged insect; 1 mm long; gray
- Sucks blood.
- Burrows into and lives under the scales of the feet, causing lifting and separation from underlying skin. Results in swelling, tenderness, scabbing, and deformity. Related joint problems may occur.
- Coating the entire leg shaft with petroleum jelly or vegetable, mineral, or linseed oil every two days may help smooth and moisturize scales.
- Dust or spray birds, coop, and roosts with carbaryl products such as Sevin, carefully following the instructions on the label. Always consult with your veterinarian before using such products.
- Ivermectin pour-on (also called spot-on) is a systemic agent used to control both internal and external parasites.

A nice hatch of Easter Egger chicks has just graduated from their brooder box to this coop with outdoor access. Moving your chickens to a new environment can heighten their stress level. This is a time to keep an extra close watch on behavior.

states and in California. A respiratory infection caused by *Haemophilus paragallinarum* bacteria, coryza often manifests in combination with other chronic respiratory diseases. While contagious, it usually isn't fatal.

Infectious Bronchitis: Infectious bronchitis is a common, fast-spreading, and highly contagious respiratory disease caused by several strains of coronavirus. While chicks often succumb to the infection, adult chickens generally survive,

but they remain carriers for life.

Fowl Pox: There are two types of fowl pox: wet, sometimes called Avian pox, and dry, also known as fowl diphtheria (neither is related to human chicken pox). Caused by the same virus and transmitted in the same manner, the former is primarily a skin disease while the latter affects both the skin and the respiratory tract.

Infectious Laryngotracheitis: Infectious laryngotracheitis (sometimes called

Avian diptheria) is a slow-spreading, but serious, upper respiratory tract infection caused by a herpes virus. It manifests worldwide and commonly afflicts laying hens during the winter months.

Marek's Disease: Marek's disease is a global scourge; it kills more chickens worldwide than any other disease. Marek's symptoms vary according to its victims' ages but it often culminates in sudden death. Caused by six different herpes viruses, it is virulently contagious. Chicks vaccinated at one day of age are usually immune for life.

Newcastle Disease: Newcastle disease manifests in many forms and in varying degrees of seriousness. It's not generally fatal, and survivors gain immunity for life. Caused by a paramyxovirus, Newcastle can trigger minor eye infections in humans who handle the live vaccine or infected chickens.

Picking and Cannibalism

Although not diseases, picking and cannibalism are the most vexing aspects of chicken keeping and clearly major threats to the health of your flock. On occasion, chickens literally peck each other to death. Worse, once it happens, the habit is readily established. It's important to nip this vile habit in the bud with adults and chicks alike.

Cannibalism usually takes root when one of the flock is injured. The sight of blood draws her peers because chickens peck at anything red. Unless the injured chicken is removed, the pecking escalates and she's likely to be pecked until she's dead. Although mildly injured chickens can be left in the flock and treated with commercial products such as Hot Pick to thwart further pecking, it's better to separate them until they've fully recovered. Installed in their own quarters

Molting and an aggressive rooster are both causes of this hen's (at right) damaged feathers. Relocate a hen who's been the target of a flockmate's aggression to prevent further injury.

and administered antibiotic or herbal ointments, healing will be hastened. By rescuing them, you'll likely save their lives. In general, don't leave wounded, lame, weak, undersized, odd-colored, or otherwise unusual birds in a flock of aggressive peers. To preserve their lives, move them to safer quarters.

Many other factors influence cannibalism within a given flock, including overcrowding, intense lighting and heat, diet, breed-related problems, and stress.

Address the problems of overcrowding by removing some birds, moving the flock to a roomier coop, or turning them outdoors. You can also install additional waterers, feeders, and nesting boxes. Dim the lights, install fans, and create more shade to prevent the overly high activity level that results from bright lighting and the edginess caused by sizzling, steamy heat. If you suspect that diet is the culprit in negative behavior, try different combinations until you find something that works. For instance, confined, solely scratch-fed chickens sometimes don't receive sufficient nutrients from their diet. If they begin pecking one another, switching to commercial feed or a commercial-feed or scratch blend sometimes helps. Conversely, chickens fed a strictly commercial diet sometimes peck out of boredom.

Strewing scratch grains, garden greenery, or acceptable table scraps adds dietary variety, and scratching and nibbling at these goodies gives idle chickens something to do. Flighty, nervous birds such as Leghorns, Minorcas, and many other Mediterranean breeds are more likely to peck; circumvent this problem by ruling them out for your farm. Finally, and most fundamentally, avoid situations that will leave your birds stressed, such as rough handling, temperature extremes, and abrupt changes in routine.

We kept a close eye on these newly hatched Red Jungle Fowl chicks. Put an immediate stop to picking among your young birds—the habit can quickly lead to cannibalism.

First-Aid Kit

Be ready for inevitable avian emergencies by assembling a chicken first-aid kit. You will need:

- A sturdy container. Supplies for a small flock fit neatly in a large fishing tackle box or a lidded bucket. Keep the container stocked; if you use something, replace the remainder when

Our first-aid kit contains all that we would need to treat an injured bird.

you're finished or replenish the used stock with a new supply. Store the kit where you can find it quickly. Write your veterinarian's phone number on the lid using permanent marker (program the number into your cell phone and post it by the house phone, too).
- A flashlight with a strong beam. Tuck it to a zip-lock bag with an extra set of batteries and a fresh bulb.

Into a second zip-lock bag, place injury treatment materials such as:
- First aid tape in several widths
- Stretch gauze
- Gauze pads
- Cotton balls
- Wooden Popsicle sticks to use for splints

To prevent spills and messes, cache medicinals in a third zip-lock bag:
- VetRX Veterinary Remedy for Poultry (for respiratory distress)
- Anti-pick spray or ointment
- Wound powder, herbal wound dressing, antiseptic spray or ointment
- Saline solution for cleaning wounds and injured eyes
- Holistic chicken keepers might add Bach Rescue Remedy (or Cream)
- Any other medicines your veterinarian recommends

A final zip-lock bag contains hardware:
- Scissors
- Dog toenail clippers (for snipping nails and beak tips)
- One or more 5-cubic-centimeter or larger catheter-tip disposable syringes (for feeding and watering debilitated chickens)

Add a pet carrier for transporting sick chickens to their vet and a quarantine cage, and you'll be set for most any emergency. Put it together now, before you need it. You (and your chickens) will be glad you did.

Major Chicken Maladies

Disease	Symptoms
Avian Influenza (Virus)	Mild form: coughing, sneezing, decline in egg production
Fowl Cholera (Bacterium)	Systemic form: chronic respiratory infection, high mortality
Infectious Coryza (Bacterium)	Green diarrhea, darkened head and comb, swollen feet, paralysis, swollen wattles, listlessness, high mortality
Infectious Bronchitis (Virus)	Watery eyes, putrid nasal discharge, swollen face and wattles, sneezing
Fowl Pox (Virus)	Coughing, sneezing, gasping, nasal discharge, respiratory distress, depression, marked drop in egg production
Infectious Laryngotracheitis (Virus)	Dry form: brownish-yellow lesions on the unfeathered skin of the head, neck, legs, and feet. Wet form: labored breathing, lesions in the mucous membranes of the mouth, tongue, upper digestive tract, and respiratory tract. Reduced egg production; low mortality
Marek's Disease (Virus)	Partial paralysis, blindness, wasting, tumors

Prevention	Comments
Vaccination	A reportable disease spread by people, birds, and flies
Vaccination (but not until cholera has been diagnosed in the flock; even so, vaccination may not be effective); proper sanitation and rodent and predator control; proper disposal of dead birds	Caused by secretions from carrier birds, cannibalism of dead birds, contaminated water, feed, equipment, or clothing
Don't indiscriminately introduce outside birds into your flock; if possible, raise replacements from day-old chicks	Spread by direct bird-to-bird contact and contaminated feed and water; often introduced to the flock by seemingly healthy carrier birds
Vaccination, proper sanitation, quarantine of all new birds	Caused by inhaled airborne respiratory droplets and contaminated food and water; often introduced when infected birds are added to a flock
Vaccination	Spread by mosquitoes and ingestion of sloughed scabs
Vaccination	Spread mainly via infected birds' droppings
Vaccination shortly after hatching	Caused by inhalation of virus-laden feather follicle dander. Spread mainly via infected birds' droppings

chapter NINE

Farm Fresh Eggs and Finger Lickin' Chicken

Whether you keep chickens for pleasure or for profit, chances are you use or sell their yummy eggs. Raising chickens for meat is a sure way to know exactly what does—and does not—go into your bird before it reaches your table; you'll know the donors were handled humanely and exactly how their eggs or flesh was processed. Homegrown poultry and eggs are infinitely fresher and tastier than anything you can buy in a store, and producing wholesome, farm-fresh meat and eggs is cost-effective and relatively easy. Want to try? Here's how.

Getting the Best Eggs

Getting the best eggs for your purposes means choosing the right breed for your purposes. You then must make sure you're providing the hens with the best possible diet and living conditions. You need to understand what factors affect egg laying and how to circumvent possible problems.

The Right Hen for the Job

Among production breeds, Leghorns are the queens of the henhouse; they lay early and often. Their compact, wiry bodies put everything they have into laying eggs. Yet they're not many small-flock keepers' first choice. Leghorns are noisy and flighty, and if you eat your spent hens, there's not much stewed chicken on those bones when a Leghorn's laying days are through.

A slightly more substantial layer is the Red Star sex-link hen, a reddish brown bird accented with flecks of white. The product of a Rhode Island Red rooster and a Leghorn hen, the Red Star sex-link lays handsome brown eggs and is far less flaky than a Leghorn. Another benefit: since pullets are red and cockerels are white, you can tell hatchlings apart—hence the "sex-link" part of the name.

Her close cousin is the gold-accented Black Star sex-link hen (cockerels are black with white barring), whose mama was a Plymouth Rock and daddy a Rhode Island Red.

The Black Star lays bigger eggs, though slightly fewer, than the Red Star. Like the Red Star, she's a fairly easygoing bird.

Old-time layers and dual-purpose chickens, such as Brahmas, Dominiques, Cochins, and Wyandottes, have a place in today's henhouse, too. They begin laying later and don't lay as many eggs, but they keep it up longer than production-breed hens.

Bantams for eggs? Sure, why not? Bantam eggs are tiny—smaller than peewee eggs from the store! But fanciers claim they're the tastiest of all hens' eggs, and bantam layers require very little feed and space.

Though color is purely a matter of aesthetics, not taste, if you plan to sell extra eggs when you start your flock, choose a breed that lays the color most popular in your locale (see "Pick a Color," opposite).

Egg-Laying Timetable

When your home-raised pullets are six weeks old and ready to leave the brooder, move them to their own safe haven, away from aggressive older chickens. Switch their feed to a 15–16 percent protein grower ration and optional supplements such as "big girl" scratch and greens. Don't forget to set out a free-choice grit container.

Around twenty weeks of age, upgrade to a 16–18 percent layer ration and add calcium-boosting, free-choice oyster shell alongside their grit. Never let feeders or waterers run dry. Keeping fresh, pure drinking water in front of your hens is a must! Even a few hours without water affects their lay. If your hens are super-layers, such as Black or Red Star sex-links, or from fast-maturing production Leghorn strains, they'll begin laying between twenty and twenty-four weeks of age.

A pullet's first eggs will be teensy treasures. New layers rarely grasp the concept of nest boxes, so you'll find eggs wherever they land. Tuck an artificial egg, such as a wooden or marble one, a sand-filled plastic Easter egg, or a golf

Research chicken breeds to select a hen who will lay eggs the color of your choice.

Pick a Color

Brown eggs
Aseel, Australorp, Sex-Links (both Red and Black), Brahma, Buckeye, Chantecler, Cochin, Cornish, Delaware, Dominique, Faverolle, Java, Jersey Giant, Langshan, Malay, Marans, Naked Neck, New Hampshire, Orpington, Plymouth Rock, Rhode Island Red, Welsumer, Wyandotte

White eggs
Campine, Crevecoeur, Dorking, Houdan, La Fleche, Lamona, Leghorn, Minorca, Polish, Redcap, Buttercup, Silkie, Sultan, White Faced Black Spanish, Yokohama

Tinted eggs (not quite white)
Ancona, Campine, Catalina, Hamburg, Lakenvelder, Modern Game, Old English Game, Sumatra

Colored eggs
Araucana, Ameraucana, Easter Egger

ball, in each nest. Pretty soon your hens will understand and begin laying in the nesting boxes.

By week thirty-two, most hens will be up to form. They'll continue laying full bore for at least two years and can continue laying for up to twelve years. As they age, hens' eggs will increase in size but decrease in number.

Remember that hens don't lay while brooding or raising chickies, and they may stop laying as winter days grow short. Hens require fourteen hours of daylight to keep producing, so in northern climes, lighting the henhouse is an absolute must. Add extra hours of light pre-sunrise so your chickens will naturally go to roost at dusk. Use a timer so you don't forget; it's important to be consistent. Just one or two days without additional lighting can throw their production out of whack. Or give your hens a winter break; they'll ultimately last longer if you do.

Each year, your girls will molt; they'll shed and regrow their feathers a few at a time. Molting generally begins as summer winds down and extends for twelve to eighteen weeks, through early fall. A fast molt is a fine chicken trait indeed because egg production slows or ceases as Ms. Hen molts. The sooner she's finished, the sooner she'll lay.

Stressed hens lay fewer eggs. A whole passel of things can stress chickens: extreme heat or cold, fright (don't let kids, pets, or predators harass them), illness, parasites, adding new chickens to the flock, or taking away familiar friends. Business as usual keeps stress down and is good for

Layer Requirements

FEED

Pullets, 6–20 weeks Layers, 20 weeks and up Breeders, 20 weeks and up	15%–16% protein grower ration 16%–18% protein layer ration 18% protein breeder or layer ration
Necessary supplements	Oyster shell for additional calcium; grit (not necessary if confined and eating commercial rations only)
Optional dietary supplements (using will upset the commercial feed's nutritional balance and may reduce egg production)	Scratch grains, greens, most garden produce and fruit (no potato skins, no avocados), bugs, and other foraged goodies

HOUSING

Indoor floor space for free-range chickens and chickens with adequate outdoor runs and indoor roosts	Heavy breeds: 4 sq. ft. Light breeds: 3 sq. ft. Bantams: 2 sq. ft.
Outdoor run requirements for above	Heavy breeds: 10 sq. ft. Light breeds: 8 sq. ft. Bantams: 5 sq. ft.
Indoor floor space for confined birds without access to outdoor runs	Heavy breeds: 10 sq. ft. Light breeds: 8 sq. ft. Bantams: 5 sq. ft.
Starter Roosts (4 in. from floor and 12–14 in. between rails; move lower rail higher as birds mature)	Heavy breeds 6–20 weeks old: 6–8 in. of space; light breeds 4–18 weeks old: 4–6 in. of space; bantams 4–18 weeks old: 3–4 in. of space
Permanent Roosts	Heavy breeds: 10 in. of space, 18 in. from floor, and 16 in. between rails; light breeds: 8 in. of space, 24 in. from floor, and 14 in. between rails; bantams: 6 in. of space, 24 in. from floor, and 12 in. between rails
Nesting boxes	1 for every 4–5 hens (1 for every 3 hens in flocks of 12 or less); bed deeply; change litter every week
Feeders	1 standard hanging tube-style feeder per 25 birds or 4 in. of trough space per hen
Waterers	1 inch waterer space per hen, providing at least 2 gal. of water per 25 birds
Lighting	One 25- to 40-watt bulb per 40 sq. ft. of floor space, placed at ceiling height above feeders/waterers; provide 14–16 hours of light per day year-round

This light Brahma vacates the row of open-topped nesting boxes, leaving a fresh-laid egg for our breakfast.

laying. Strive for peace and serenity in the henhouse if you love fresh eggs.

Who's Been Eating My Eggs?

Everyone loves fresh eggs. Even hens. Egg eating begins innocently enough, when an egg is accidentally cracked or shattered and a curious hen takes a nibble. Mmm-mmm, good! She keeps her eyes peeled for more golden goodness and when she tucks in, her sisters' curiosity is piqued. They sample, too. Yummy! One fine day, someone notices that if you peck really hard, you can sometimes serve yourself. Pretty soon, your flock is eating more eggs than you are.

What to do?

- Revamp your coop's nesting area. Provide more nests for the flock so there is less traffic and, ultimately, fewer broken eggs. After all, a single broken egg can trigger this hard-to-zap habit.
- Relocate nesting boxes away from the fast lane. Install them at least 24 inches from the floor in a secluded corner of the coop.
- Keep plenty of clean, cushiony litter in each box. Protective padding saves many an egg.
- Ban broodies from the henhouse. They're happier off by themselves, and the nest they're setting in is one less for the rest of the hens to use, resulting in higher traffic in the remaining cubicles.
- Strive for stronger shells. Feed high-calcium commercial layer ration with oyster shell served up free-choice.
- Pulverize egg shells to feed to your hens. They're a dandy source of calcium, and getting used to the

taste won't give a hen that "Ah-ha!" moment when she realizes she's dining on egg.

- Stressed chickens pick. Keep everything in your hens' environment as low-key as you possibly can. Avoid changes in their daily routine and never let them run out of fresh feed and water. Don't introduce new chickens to an established laying flock, which triggers changes in social order. Keep your chickens reasonably cool in July and warm in February. Absolutely avoid overcrowding and always handle laying hens in a quiet, compassionate manner.

- Don't push hens roughly out of nest boxes when collecting eggs. If a hen breaks eggs as she hastily retreats, clean up the mess right away.

- Identify culprits and cull them to pet homes—or even the stew pot. You'll know them by the dried yolk remnants decorating their beaks and heads.

- Don't assume your hens are noshing all those eggs. Predators such as skunks, opossums, and the occasional snake fancy chicken eggs, too.

Tasty Chicken

If you want to grow a lot of tasty chickens in a short time, start with meat-breed chicks. Chicken-raising newbies who spring for low-priced Leghorn cockerels will be disappointed. Bred for laying eggs, not for making meat, light breed chickens guzzle twice the feed and never flesh out to prime eating size.

Super or Dual-Purpose Birds

Super-broilers convert feed to flesh at lightning speed. They take eight to twelve weeks from hatch to slaughter at

Egg yolks and crushed shells are tasty treats to a chicken. Promptly remove eggs from the henhouse, or you may find your birds eating more eggs than you do!

Great-Tasting Eggs

Follow these tips for clean, great–tasting eggs:

- Use plenty of cushy nesting-box litter and check it often. Remove muck promptly. Dump everything out and replace it with fresh, fluffy litter at least once a month.

- Collect eggs first thing in the morning and at least once or twice more during the day. The longer the eggs stay in the nests, the more likely they'll become mud-smeared, splotted upon, or cracked. For the same reason, supply plenty of nesting boxes. Cutting down on traffic really helps the eggs.

- Collect eggs in a natural fiber or coated wire basket (not the calf's drained nursing bucket or the horse's sweet feed scoop). To prevent breakage, don't stack them more than five layers deep.

- Don't clean eggs if they don't need it. Carefully chase minor spots with fine grit sandpaper. When you must wash soiled eggs, do it as soon as you can, before the eggs have cooled. Cooling causes shells to contract and suck dirt and bacteria into their pores.

- Use water that is 10 degrees warmer than the eggs themselves. This causes their contents to swell and shove surface dirt out and away from the shell pores. Gently scrub them in plain water or mild egg-cleaning detergent (available from poultry-supply retailers). Never soak eggs, especially in cool water; water contaminants can be absorbed through the shells.

- Dry washed eggs before storing them.

- Refrigerate eggs in cartons, large-end up. To preserve quality (and prevent development in fertile eggs), get them in the fridge as soon as you can. Date the cartons and rotate your stock, using up older cartons first.

- Eggs absorb odors from strongly scented foods such as fish, garlic, onions, cantaloupe, and apples. Try not to store eggs with these items.

Dual-purpose breeds such as this silver-laced Wyandotte rooster produce good meat, but mature more slowly than super-broilers such as Cornish Rocks.

No Names

Years ago, my husband and I raised chickens for meat. Just one batch, as it turned out. Our mistake: we ignored an age-old adage—don't give your food names. We made pets out of our boisterous young cockerels. When processing day rolled around, we could only bring ourselves to kill a few; we gave the rest away. Two weeks later, we were vegetarians. Sage advice from she who knows: don't give your broiler chicks names!

4 to 5 pounds live weight, or they can be slaughtered earlier (at five to six weeks as Cornish Game Hens) or later (at twelve to twenty weeks and 6 to 8 pounds live weight) if the raiser prefers birds of a different size.

Old-fashioned dual-purpose breed chickens make for delicious eating, too, although they mature slower and demand more feed for each pound of weight they pack onto their sturdy frames. Old standbys such as Rhode Island Reds, White Rocks, and New Hampshires need twelve to sixteen weeks to grow to broiler size, but their slower-growing ways spare them the health and structural problems that super-broilers experience during their far-shorter lives. Old-fashioned breeds are more flavorful, too. In addition, if you're licensed to sell dressed birds, meat from most old breeds can be labeled "Heritage Chicken" (see chapter 4).

Most super-broilers and popular dual-purpose breeds are yellow-skinned, white-feathered chickens, simply because that's the kind of bird Americans prefer. Brown and black semi-super-broilers can be ordered from large hatcheries such as Stromberg's and McMurray's (see Resources). They're more active than white Cornish-Rocks and they bloom a smidge more slowly, but like dual-purpose breed broilers, they're less susceptible to the structural problems that plague their faster-maturing kin.

Broiler's Timetable and Requirements

Whichever type you prefer, you must feed your meat peeps 20–22 percent protein broiler starter. Lower protein products simply won't do. Check with your feed store before chicks arrive because they may not keep broiler feeds in stock. Figure one hundred pounds of starter per twenty-five chicks.

Broilers and Roasters
Broiler / Roaster Requirements

FEEDING	
Broiler chicks, first 6 weeks	20%–22% protein broiler starter ration
Broilers, 6 weeks to slaughter	18%–20% broiler finishing ration
Necessary supplements	Grit (not necessary if confined and eating commercial rations only)
Optional dietary supplements (feeding will upset commercial feeds' nutritional balance and birds will mature more slowly)	Scratch grains, greens, most garden produce and fruit (no potato skins, no avocados), bugs and other foraged goodies
Optional drinking water additives	Vitamin-electrolyte supplement
HOUSING	
Indoor floor space	6 to 10 weeks of age: 2 sq. ft.; 10 weeks to slaughter: 3 sq. ft.
Roosts	None
Feeders	1 standard hanging tube-style feeder per 20 birds or 4 in. of trough space per bird
Waterers	1 in. of waterer space per bird, providing at least 2 gal. of water per 25 birds

This barnyard rooster is mostly Old English Game, a good forager and, therefore, good in free-range situations.

When your chicks leave the brooder at roughly six weeks old, switch to broiler finisher. They'll remain on this ration until they're slaughtered.

Most folks raise broilers indoors, allowing 2 square feet of floor space between 6 and 10 weeks of age, then 3 square feet until slaughter. Continuous, ultra-low lighting encourages nighttime noshing, which is especially important during sizzling summer climates. Some raisers limit feed intake, others keep feed in front of their birds all the time. Whichever method you choose, make sure that plenty of clean, cool water is available. Adding a vitamin-electrolyte product to their drinking water is a wise idea, too.

Broiler Don'ts

Unfortunately, astronomical weight gain comes at a hefty price. Cornish-Rock broilers' broad-chested, meaty bodies mature faster than their skeletal structures can support them. Crippled legs and crooked breastbones are the norm.

Lame birds crouch to ease pain, and they develop breast blisters. Super-broilers are also prone to heart attacks. For best results with swift-maturing meat birds, heed these ploys:

- Stir litter often; keep it fluffy and dry. Crouching on hard-packed litter irritates broilers' keel bones. It's a major cause of breast blisters.
- Don't give meat chickens roosts; as

they hop to the floor after roosting, they'll damage their legs. Pressing against these hard objects causes breast blisters, too. Remove anything in their environment that they could leap up on: rocks, boards, doors, ledges, protrusions of any sort.

- Don't capture or carry meat fowl by their legs.
- Don't startle or chase these injury-prone fowl. Keep surprises down to a minimum. Broiler stampedes equal torn muscles, slipped joints, and heart attacks.
- Remove lame chickens from the flock. Lodge them in separate quarters with easy access to food and water. Many will recover within a week. To prevent needless suffering, cull any birds who don't.

Making Meat Down on the Farm

Butcher all of your broiler chickens at the same time or create a continuous home supply by starting a new brood when the first is four weeks old, then slaughtering one-fourth of the older group at intervals of seven, eight, nine, and ten weeks of age. Process roasters at 6 to 8 pounds of live weight. Don't carry them much beyond that. The older they get, the less efficiently they convert feed to meat.

Adding vitamin and electrolyte powdered supplement to your new chicks' waterer at the manufacturer's dosage recommendation can help prevent leg weakness in certain broiler breeds.

If you've never processed chickens before, have someone show you how. The process isn't tricky, but it should be done with precision. Second-best, print or download a heavily illustrated university publication and follow its instructions to the letter.

Keep It Clean

To quash disease and avert parasite problems, always strip, scrub, and thoroughly disinfect your broiler quarters between batches of birds.

If you do, and you raise your chickens indoors, you usually needn't vaccinate or deworm them, especially if you feed them medicated commercial rations.

chapter TEN

Bucks for Clucks

Y ou don't need a huge flock of chickens to earn a tidy amount of egg money with your feathered friends. In fact, there are several tried-and-true ways to make money with chickens. Here are a few to consider.

Sell Fresh Eggs

It seems the logical thing to sell excess eggs to family and friends or maybe from a roadside stand. But first, investigate the egg laws in your community, county, and state to make sure that selling eggs is legal and, if it is, to learn how you're required to store and market your farm-fresh eggs.

Many states' egg laws, also called egg statutes, are available online, but if yours aren't, your county extension agent is the person to see. He or she can put you in touch with your state's egg-sales regulatory program. Laws vary widely from state to state, and some states impose stiff penalties for simple infractions that you may not even be aware of, so don't omit the important step of familiarizing yourself with the regulations.

Consider my own state's rules. The Arkansas Egg Marketing Act states that people who own less than 200 hens can sell eggs directly from their farms, providing that the following requirements are met:

1. Eggs are washed and clean;
2. Eggs are prepackaged and identified as ungraded with the name and address of the producer;
3. Used cartons are not used unless all brand markings and other identification is obliterated; and
4. Eggs are refrigerated and maintained at a temperature of forty-five degrees Fahrenheit (45°F) or below.

Small-producer laws such as these usually apply to selling eggs from your farm. Selling at farmers' markets or flea markets, through community-supported agriculture

(CSA), or directly to restaurants or grocery outlets—even natural food stores—is generally another story; in most states, you must be licensed, and more stringent laws apply.

Eggs for Direct Sales

When you market eggs, go the extra mile to produce a really fresh product. Keep those nest boxes extremely tidy and gather eggs at least twice a day (more often is better, of course) to prevent soiling and get them under refrigeration as quickly as possible.

Choose productive breeds that lay the color of eggs—brown or white—that buyers in your locality prefer. And consider a specialty product such as all-blue eggs from Araucana hens (they also happen to be on the ALBC Heritage Chicken list), free-range eggs, or organic eggs. However, if you call your eggs organic,

What about Those Cartons?

Nearly every state's egg laws prohibit the reuse of commercial egg cartons unless all previous markings are fully eliminated. Most, however, allow you to reuse your own cartons if customers bring them back to be refilled.

Salvage commercial cartons, if you like, by covering original markings using a product such as Diagraph Quickspray Blockout Ink (see Resources), an aerosol or brush-on cover-up that comes in tan or white. It doesn't, however, look professional, so if you're promoting your eggs as a quality product (and you should), consider starting with new, unmarked cartons. Eggcartons.com and similar companies (see Resources) sell plain or custom-imprinted paper, foam, and plastic egg cartons at discount prices. Mail-order chick hatcheries such as Murray MacMurray (see Resources) sell them, too. Even farm stores sometimes keep new egg cartons in stock. Paper (or pulp) cartons generally cost less, and they're Earth-friendly, so many consumers prefer them. Be sure to order some six-egg cartons as well as twelve-egg models; you can market six eggs at a slightly higher cost per egg, and they're convenient for clients who don't consume a lot of eggs.

In virtually every state, you must mark your cartons with your name and contact information. If you're selling specialty eggs (free-range, organic, ALBC Heritage), that should be prominently noted on each carton, too.

Make a good impression by printing an attractive label to affix to blank egg cartons. If your hens are Rhode Island Reds, add that information and a picture of a Rhode Island Red. Add another label with hints on the perfect boiled egg or a recipe for luscious French toast. Consumers like this type of information, and it makes your product stand out among the rest.

The "Honor-System" Roadside Stand

In the olden days, an egg producer would set up a card table by the roadway, place several dozen eggs in cartons and an empty coffee can with a slit in the plastic lid on top of the table, and tack up a sign that said "Farm Fresh Eggs, 50 cents a dozen." Customers dropped by, selected a carton of eggs, and dropped their coins in the slot.

Eggs cost more these days, but this system still works surprisingly well in some locales (honor-system, self-serve selling is alive and well, for instance, in the southern Ozarks) with two caveats: first, while most people are as honest as their parents and grandparents were, occasionally someone is bound to trash the till. If you engage in self-serve selling,

Even today, farmers in small communities trust their patrons to leave payment for the eggs they take.

make it a point to empty the money can at least several times a day. Second, cool those eggs! Run a cord to a dorm refrigerator set up under your card table or at least supply a cooler and ice, replenished as needed. We should be much wiser about food poisoning nowadays.

If you think blue eggs like this one will sell better, purchase an Araucana hen.

eggs accordingly. Visit at least three mainline grocery stores to see what eggs are bringing, then stop by a natural-foods store and take note of its egg prices Ask the clerk at the natural-foods store what makes his eggs worth the extra price. If your eggs share those qualities or even better, price them in the health food store's range or slightly higher.

Marketing Eggs

Plan to sell your eggs from your farm unless you've obtained the licensing required to sell through other outlets. Word of mouth sells a lot of eggs, so tell your family and friends that you're in business and let them spread the word. Look ahead to the end of this chapter for more ideas on how to promote your farm-based chicken business.

Sell Hatching Eggs

Another good way to earn a bit of egg money is by selling hatching eggs, especially if you keep rare or unusual purebreds that other chicken fanciers would like to own. Hatching eggs and an incubator are the ideal solution for people who want to raise

be sure that they really are. Stiff penalties apply to marketing non-organic foods as organic. Visit the National Center for Appropriate Technology website (see Resources) and read "Organic Poultry Production in the United States" to see what organic entails.

Even if you don't wash eggs for home use, you will be required by most state egg laws to wash eggs that you are going to sell. Wash eggs carefully (see chapter 9), then place them in clean cartons and pop them in the refrigerator as soon as you can.

Pricing Eggs

Don't underprice your eggs. According to figures published by the United States Department of Agriculture, the average price for one dozen organic eggs was between $2.55 and $3.16 per carton in April 2010. Customers expect to pay more for niche products, so price your

Did You Know?

The American Poultry Association says that hens of the best laying breeds, such as Leghorns and their sisters from the Mediterranean class, lay between 250 and 280 eggs per year. More conservative sources claim that a good hen lays about 180 eggs per year.

rare-breed chicks but not the twenty-five peeps needed for live shipment.

A good place to sell hatching eggs is on eBay (there are more than 1,000 hatching-egg auctions up for bids as I write this; among them are ALBC Critical-listed Heritage breeds and ultra-rare breeds such as Seramas) or Eggbid, the eBay of the poultry world (see Resources). Ads in the chicken fanciers' magazines listed at the back of this book bring results, as do classified ads in magazines such as *Hobby Farms* and *Chickens*. Don't forget freebie ads in PennySaver-type newspapers.

You must be ultra-conscientious when selecting eggs to sell for hatching purposes. Collect eggs at least four times a day, sort them for size and conformity, and store them in a room in neither cold nor hot conditions, away from drafts and sunlight. Turn them at least twice a day. Eggs should be no more than three days old when you ship them. Hatching eggs should be clean but unwashed, fertile, and well shaped (no wrinkled or misshapen eggs). While it's widely understood that a seller assumes no responsibility for eggs once they're shipped, send the best eggs you possibly can.

There are many secure ways to ship hatching eggs, but this is a good one (search "ship hatching eggs" online for additional ideas):

- Obtain free Priority Mail or Express Mail shipping boxes from your post office. The size 4 Priority box (6×6×6 inches) is great for sending a few eggs; size 7 (12×12×8 inches) is perfect for shipping a full dozen or more.
- Use lots of bubble wrap and packing peanuts. Cut strips of bubble wrap about 6 inches wide and 12 inches long to ship full-size eggs. Wrap each egg around its sides and tape it, then follow with a second piece of bubble wrap and tape. You can still see both ends of the egg at this point. Wrap a third piece of bubble wrap around the egg lengthwise to protect the ends, then set that egg aside.
- Place a layer of packing peanuts in the bottom of your box; then line it with several layers of bubble wrap and place the eggs close together inside. Pad the sides as needed, leaving several inches at the top in case the shipper sets something heavy atop your box. Fill that space with packing peanuts, too.
- Close the box and check to make sure that the sides don't bulge (if they do, eggs may crack when the box is stacked with others at the post office or in transit), then shake the box lightly to be sure that there is no movement inside. Finally, seal all of the edges using wide packaging tape.
- Mark all sides of the box "Fragile—Hatching Eggs." Also write the buyer's phone number on the box in large letters, along with a note requesting that postal workers call the buyer when the box arrives at the post office.
- Ideally, you should send the box on Monday morning. Never send one on Thursday, Friday, or Saturday, when it might be left in a cold or hot truck or mail-handling facility over the weekend.

Sell Live Chickens

You can also incubate extra chicks to sell, though you should talk to your county extension agent before you do. Many states consider anyone who sells live chicks a hatchery, requiring the seller to obtain a license and show proof that his or her flocks have been tested for pullorum

Who doesn't love a soft little chick? It won't be hard to sell the cuties, but be responsible about who you sell them to. Children who want chicks don't necessarily want chickens. Make sure your grown-up chicks don't end up in shelters or worse: released into "the wild."

What about Dressed Chickens?

While it's perfectly possible to sell dressed birds, don't do it before discussing the legalities with your county extension agent, who can put you in touch with the organization that oversees the sale of home-processed meat in your state. Be prepared to jump through a lot of hoops. If you want to sell meat chickens, it's easiest to sell them as live birds and let buyers take the process from there.

disease (a severe diarrheal disease of young poultry).

Swap meets are popular places to sell chicks, but they also present many opportunities to expose unsold chicks to disease. It's better to place a classified ad.

Another option is to raise those chicks and sell them as ready-to-lay pullets or as young laying hens. Many people are anxious to get into chickens but don't want the bother of raising chicks. In most places, pullets and young laying hens of everyday breeds fetch between $15 and $30 per chicken.

If you raise show-quality chickens or rare-breed fowl, you may choose to sell them to breeders far away from your locale. You can do this, weather willing (chickens can't be shipped during extremely hot or cold weather), although it isn't cheap (figure about $35 to $45 to ship an adult heavy-breed chicken).

Breeders frequently ship birds from pullorum-free flocks via USPS Express Mail, and you'll need an approved bird-shipping carton to do so. This is a sturdy cardboard box with a tapered top and holes punched in it, designed to safely transport one or more 6-ounce or larger birds. The tapered top ensures that the box won't have others stacked atop it while the holes allow for adequate airflow. Suppliers, such as Horizon Micro-Environments (see Resources), and some major hatcheries carry these boxes. The Horizon Micro-Environments boxes are good ones. Sized to ship up to two full-size or four adult bantam chickens, these boxes have mesh inserts under the pre-perforated breathing holes and even provide space for the buyer's mailing address.

Insurance guarantees that your birds will reach their destination in the quoted time frame, but it does not guarantee that they'll be alive, so it's up to you to take pains to ensure their safe shipment. Make certain that they are healthy, well fed, and well hydrated prior to shipping day. Place a thick layer of absorbent material in the bottom of the box (coarsely chopped wheat straw works well) and supply moist edibles such as orange or apple slices, fresh pineapple slices, or watermelon rind to help hydrate the birds during shipment. As with hatching eggs, write the buyer's phone number on the box and request that the buyer be called as soon as the birds arrive.

Find out when mail pickup occurs, and take boxed birds to your post office no more than an hour before the truck is scheduled to arrive. Allow time to fill out the Express Mail label, or pick it up in advance and take it to the post office with you.

Save your copy of the shipping form and call or e-mail the tracking number to the buyer. Be sure that the buyer has your

These chicks are being packed up to be sent to their new home. Make sure your chickens have everything they need for the journey to prevent injury and illness.

Weird Eggs!

Normal eggs are graded in six sizes from 1.25-ounce peewees to 2.5-ounce jumbos. The size of egg that a hen lays depends on her breed and age (eggs typically get bigger as the hen ages). It takes a hen about twenty hours to form an egg, start to finish. Colored shells are created when pigments are added during their formation. Most eggs have one rounded end and one more pointed end. Barring accidents, each hen lays her own distinctive shape and size of egg day after day, so you can tell which egg in a basketful is hers. But sometimes things don't go according to Nature's usual plan. Then you get weird eggs.

- Wind eggs, sometimes called "cock eggs" or "dwarf eggs," have no yolk. They're usually laid by pullets in early production.
- Double yolks occur when one yolk moves too slowly and is joined by a second yolk before the shell is formed. Usually heavy-breed hens lay these eggs, though pullets sometimes lay them, too. According to *Guinness World Records*, the current record-holding largest egg had five yolks, and the heaviest egg on record had two yolks and a double shell!
- Double-shelled eggs are created when a ready-to-be-laid egg reverses direction and second layers of albumen and eggshell form around it. They are so rare that you may never see a double-shelled egg.
- Wrinkled or oddly shaped eggs, however, are fairly common. These occur when a hen's ovary releases two yolks uncommonly close together, causing them to travel through her oviduct in close proximity. The first yolk gets a normal-looking shell, but the second shell will be thin, wrinkled, and flat toward the pointed end.

You can sell attractive chicken feathers to fishing-fly tiers.

phone number, and ask him or her to call or e-mail you when the birds arrive.

Sell Feathers

Fishing-fly tiers and craftspeople treasure the natural, undyed saddle, hackle, and tail feathers of some breeds and varieties of roosters. Fly tiers often prefer feathers from bantam versions because of their smaller, more delicate feathers.

To harvest the feathers for sale, prepare capes from roosters that you slaughter to eat. A cape is a dried skin that comprises the neck, back, and sides of the bird. Remove it in one piece and tack it onto a hard surface, skin-side up. Scrape all of the fat and meat from the skin, then liberally dust the skin with powdered borax (find it in the cleaning section at the supermarket). Let the skin sit for a couple of weeks, scraping off any moist areas and reapplying borax as needed. When a cape is fully dry, store it in its own roomy zipper-seal bag with a little extra borax and a few whole cloves (borax keeps the cape dry, and the cloves improve its overall aroma) in a dry, cool location away from marauding cats and dogs.

If there is any chance that your feathers or finished capes could have feather mites, treat them to a two-week stint in the freezer. Then, package them attractively, and sell them from your craft-show or farmers'-market booth or on eBay.

Bang Your Own Gong

It doesn't take a lot of money to promote a home business, but promote it you must. Customers are out there, but they have to know where you are and what you sell. There are a lot of simple little ways to make a big promotional splash.

Your Chicken-Biz Website

Put up a simple website. If you don't know how, major hosting services such as Yahoo! Business offer easy-to-use page templates that walk you through the process. It's simple—I use it; you can too. Here are some tips to help you create the best site:

- Web designers recommend that a web page's elements add up to single-page downloads no greater than 180 kilobytes. This precludes huge pictures and techie frou-frou. Music, animations, and huge photo downloads annoy some users and freeze older computers.
- Choose fonts with care. Make certain that they're large enough to be easily read on both Macs and PCs and in all of the standard browsers. Avoid nonstandard fonts; when in doubt, use Arial or Times New Roman.
- Don't use patterned backgrounds. Strive for clear contrast between font and background colors. Remember, light-colored copy against a dark background looks great online but doesn't print well in some browsers.
- Place your name, contact information (including your physical address), and logo on every page, and link each page back to your site map or home page.
- Add educational content—the more, the better—so visitors have a reason to revisit your site.
- Don't let your site become dated. Tweak your sales lists, upload new pictures, or add new educational items—anything to keep your content fresh. Spend time each week checking your links, especially ones that lead to outside sources. One hour a week spent on website maintenance pays big dividends in customer approval.
- Finally, people won't visit unless they know about your site, so aggressively

promote it. Add it to your e-mail signature so that it's on every e-mail you send. Use it on your business cards, brochures, and road signs. Letter it on your truck and have it emblazoned on the T-shirts that you wear out to the store. Your website is a major marketing tool for you—a modern small-business entrepreneur.

It's in the Cards

Business cards are inexpensive, and there are scores of ways to promote your business with good cards.

- Pin your cards to every bulletin board you encounter. Use push pins so that you can stack them; this encourages people to take one. Check back often to restock.
- Tuck a business card into every piece of mail you send out: personal correspondence (your cousin Joe may know someone who's hot to buy free-range eggs), invoices, and even bank and utility payments.
- Hand out cards to people you meet. Ask them to take several and pass the extras along to their family and friends.
- Keep in mind that bent or soiled cards create an impression of your business that you want to avoid. You have only one chance to make a good first impression; make it count!

Publicity Doesn't Have to Break the Bank

It's important to advertise, but you needn't spend a lot of money to do it right. Try these low-cost ploys. They work!

- Turn your car, van, or truck into a rolling billboard by affixing magnetic signs to its doors. Include your phone number, e-mail address, and website URL in the design. Don't forget the back of your vehicle; put signs on the tailgate or in the back window.

- Spring for custom bumper stickers. Everyone who follows you down the road will be exposed to your URL or your message. Dozens of online businesses offer design-online capability with no minimum order in both vinyl and magnetic styles.
- Order T-shirts, jackets, and hats printed or embroidered with your logo and business name and wear them everywhere you go. Buy extras for your customers, family, and friends (everybody loves them and they make good gifts).
- Place a large, legible road sign by your driveway. If you live off the beaten path, provide directional signs with your farm or business name on them.
- Contact your state's or county's adopt-a-highway program and offer to pick up trash or donate funds for litter cleanup along the roadway. In return for your efforts or sponsorship, you'll get a sign that displays your business name and logo on your allocated stretch of highway. Drive to the sign and take some gentle chickens with you so that you can have someone photograph your cleanup crew standing by the sign and holding chickens. Send the photo and a story about highway adoption to your local newspaper and then get ready to see your birds in print.
- Take your chickens out in public. Show them off. Take your chickens

One of the best ways to market your eggs? A simple sign by the road does the trick.

and a display booth to gatherings of all kinds. Join an animal therapy group that will allow you to bring a friendly chicken to hospitals and nursing homes to visit shut-ins. When you have fun, gain exposure, and make people happy, everyone wins!

- Give talks and perform demonstrations. Prepare an educational program for your kids' school. Civic groups need speakers for meetings and events; let it be known that you're available. Volunteer to speak at seminars. Hold an open house.

Sponsor or lead a 4-H poultry project. Anything to get people talking about your business.

- Host a Kiss-a-Chicken booth at the county fair and donate proceeds to your favorite charity. Just make sure it's your friendliest chicken!
- Read stories to kindergarten classes and kids' groups at your local library (from children's books featuring chickens, of course). Take along a living example. And it doesn't hurt to tip off your local newspaper in advance.
- Participate in e-mail lists. People buy

Don't pass up any marketing opportunity. Placing your fresh eggs in an attractive basket and in an attractive setting can definitely help boost sales.

This is an example of the blog I write for the Hobby Farms website (www.hobbyfarms.com). My goat Martok may be pictured, but the blog covers all things farm related—including chickens! Use the Internet to your advantage every way you can.

from people they know and respect, and you can advertise on many lists for free.

- Upload videos to video-sharing sites, such as YouTube. You can showcase your rare breed, cook a dish using fresh eggs, or demonstrate how to teach a show chicken to pose. End each video with a short talk about your farm, including contact information.
- Blog, adding new content every week to keep people interested.
- Write a chicken column for a newspaper for free with one stipulation: your contributor's byline must include your contact information.
- Offer free consultations to folks interested in raising chickens. Invite them to your farm and give them good, solid advice—without a sales pitch—but show them your chickens or eggs as part of the visit.
- And one final idea: buy a chicken costume, put it on, and strut your stuff while handing out business cards and posing for pictures at parades, county fairs, and other gatherings. Outrageous? You bet! But people *love* it.

chapter ELEVEN

Fun with Chickens

C an you have fun with chickens? You bet! You can enter your rooster in a crowing contest, show your chickens, or even keep pet chickens in your home.

Keep a House Chicken

Thanks to the invention of a nifty little item called the chicken diaper (see next page), many chicken fanciers keep a house pet or two. As one house-chicken advocate succinctly says, "What's the difference between a parrot and a chicken except $3,000?" Makes sense, and the chicken may favor you with eggs!

If this idea resonates with you, start by subscribing to the free Yahoo! Groups "housechickens" e-mail group (see Resources). Formed in 2001, this ultra-friendly group is home to more than 500 contributors, most of whom keep house chickens, though some do not. If there is anything you need to know about chickens, indoors or out, they will cheerfully field your questions. It's an invaluable resource if you keep pet chickens of any kind.

Which Chicken?

Any breed can make a fine house chicken, although the smaller the chicken, the less poo it generates. Silkies, Seramas, and similar bantams are perennial favorites, though some people prefer more substantial, full-size birds. If you have close neighbors, it's best to stick to hens (even a bantam rooster's crowing can be disruptive). Chicks raised in the house are a logical choice, but adults adapt to indoor living quite well. Many sick or injured chickens move indoors "just until they're well" and never go back to the coop again. One hand-raised chick that has bonded with humans instead of chickens makes a fine single house pet. However, if you start with non-bonded adults that grew up among other chickens, plan on two—then each has a friend.

If you live in town, however, it's important to check your municipal chicken-keeping ordinances before moving any chickens indoors. Some cities that allow outdoor hens stipulate that you can't legally keep them in your home.

Chicken Diapers

If you think you'd like to keep chickens in the house, investigate chicken diapers. Ruth Haldeman of the housechickens group is the queen of chicken diapers. She designed them and has custom-sewn them for clients since 2002. Visit her chicken diapers website, described by Ruth as "a clean place for chickens and their humans to sit" (see Resources) and read her chicken diapering FAQs.

Any kind of chicken with a tail knob and stiff tail feathers can wear a diaper (rumpless breeds such as purebred Araucanas and Manx Rumpys can't). Even chicks can wear diapers when they begin growing stiff tail feathers at roughly four weeks of age. Ruth crafts a stretchy, adjustable diaper for chicks, but a typical chick grows through at least three different sizes before it reaches full growth.

An adult hen's diaper typically needs changing every two to three hours, but disposable plastic liners mean you needn't change the whole thing. Some people keep house chickens in comfy indoor cages at night and when they're away from home, so diapering is a part-time thing.

Enter a Chicken-Related Contest

Chicken contests range from the sublime (like the heady competition at rooster crowing events) to the absurd (think rubber chicken flinging competitions). For some, you need a real, live

Washable chicken diapers are a solution to a messy problem with chickens that roam the home. This diaper is my own design. Although it worked, today's commercial chicken diapers are a better bet.

chicken, and for others you don't. Here are a few to consider.

Rooster Crowing Events

The National Rooster Crow (and the National Human Crow) is held in Rogue River, Oregon, on the last Saturday of June each year. The first annual event took place on May 23, 1953, when Hollerin' Harry crowed seventy-one times in thirty minutes and earned his owner $50. A second event was held September 5 of the same year, and a rooster named Beetle Baum crowed 109 times in the same span of time, netting his happy owner a cool $100 bill. Beetle Baum's record held for twenty-five years until White Lightning crowed 112 times at the 1978 event; White Lightning's record still stands.

Nowadays, rooster enthusiasts pit their birds against one another at county and state fairs across the land. Rules vary from fair to fair, but these from the Monroe County Fair in Waterloo, Iowa, are typical.

- Roosters entered in the rooster crowing contest must be entered in the Monroe County Fair poultry classes.
- Each owner may enter only one rooster.
- Handlers must remain on the side of the cages away from the judges.

Something to Crow About

Everyone knows that roosters crow at daybreak, but the truth is, they crow around the clock—even in the dead of night, if something arouses them.

Each rooster's call is unique. Some roosters crow louder than others do. Dominant roosters have higher pitched crows than those further down in the pecking order. Roosters inherit their style of crowing, so family members sound somewhat alike.

Hens occasionally crow, too. In the olden days, a crowing hen was considered an unfortunate omen. In reality, a crowing hen may be ill, aged, or taking on the leadership of an all-hen flock.

The most remarkable avian crowing machines are roosters of the longcrower breeds. These roos don't sing "cock-a-doodle-do" or any of the onomatopoeic calls with which roosters are credited in other lands. They start their calls like other roosters do, but their final note is an eerie, sustained sound that goes on and on until the rooster runs out of breath. Longcrowers of the Drenica breed of Kosovo reportedly crow for as long as sixty seconds at a time. American longcrowers crow for seven to fifteen seconds (the typical rooster's crow is two or three seconds at best). If you think you'd like to raise these breeds, visit a video-sharing site, such as YouTube, and search for the the word "longcrower." After hearing some of these birds, you may change your mind!

If you think your rooster's crow is particularly impressive, why not enter him in a local contest?

while another shows her roo a fetching hen. It's an extra-fun event enjoyed by the roosters, their handlers, and the captivated crowd. Maybe your favorite roo and you should try it, too?

Other Fair- and Festival-Type Events

The same fairs and festivals that sponsor rooster crowing contests often offer additional chicken-themed competitions. Rooster crows for humans are crowd-pleasers, particularly competitions for the younger set. Children crowing like their favorite birds—cute? You bet! Chicken calling contests are popular, too. And is there anything funnier than a rubber chicken streaking through the air? Yet lightweight rubber fowl are hard to throw. Rubber chicken flinging contests are popular events at many a county fair.

- Handlers are encouraged to use any means to stimulate the bird to crow. However, touching the rooster, by hand or with another object, will disqualify the entry. Tossing food or other items into the cage to encourage the rooster to crow is allowed, as long as direct contact is not made between the rooster and his handler.
- Scoring will be the number of times the entry crows during the thirty-minute time limit.

At most events, roosters are housed in individual cages that are kept covered to simulate nighttime. When the clock starts ticking, off come the shades. The antics handlers go through to encourage their "roos" to crow are priceless. Arm flapping, leaping from leg to leg, and clucking or crowing at the birds is *de rigueur*. A handler might bring along a rival rooster to taunt his stalwart entry

Enjoy Egg-Related Games

The egg toss is another popular county fair event. This is a team event in which partners face each other and gently toss a raw egg back and forth. After each successful toss, each partner backs up a predetermined distance; when the egg finally breaks, measurements are taken and the next team takes the floor.

However, the World Egg Throwing Federation (WETF), headquartered in Swanton, England (see Resources), takes this event to greater heights. Several disciplines are offered in the WETF rulebook, including egg target throwing, which is played in the following manner: a volunteer human target is placed 24 feet from the throw line in front of a safety net to prevent overthrow. The thrower is provided with four eggs. The thrower must warn the target that a throw is about to commence and must receive acknowledgment from the target before each throw. The thrower hurls the eggs at the target and is awarded points for strikes on specific areas. Eggs that don't break on impact with the target or safety net may be rethrown.

An even stranger WETF event is Russian egg roulette, an event in which players sit across from each other at a table. They flip a coin, and the winner decides whether to go first or second. A tray containing six specially selected eggs (five hard-boiled, one raw) is placed on the table. Each player takes turns selecting an egg and then smashing it on his or her own forehead until a player finds the raw one. The finder of the raw egg loses the game or match.

Why do they do it? According to the WETF website, "Egg Throwing has been a sport enjoyed by millions of people since early humans discovered the delight of watching a failure of another to catch a tossed egg." Makes sense to me!

Show Your Chickens

A more serious poultry pastime, especially at events sanctioned by the American Poultry Association (APA) and the American Bantam Association (ABA), is showing chickens. Poultry shows are popular county fair competitions, too.

Types of Shows

Sanctioned shows are the "big time" of chicken showing. Chickens are judged according to their breed standard in the APA *Standard of Perfection* or the American Bantam Assocation (ABA) *Bantam Standard* and shown in classes with others of the same class, breed, and variety in one of four groups, divided by age and sex: cock (a rooster of one year old or older, though some shows define a cock as being hatched before January 1 of the current year), cockerels (roosters under one year of age or hatched since January 1 of the current year), hens (females one year old or older, or hatched before January 1 of the current year), and pullets (females under one year old, hatched since January 1 of the current year).

Less formal nonsanctioned shows are held at county fairs. In these venues, chickens may be shown in purebred (also known as *exhibition chicken*) classes divided by APA class, breed, and variety or in commercial poultry classes for meat birds. Exhibition chickens are judged based on how well they adhere to what's set forth in the APA's *Standard of Perfection* for their breed.

Chickens are immensely popular projects for 4-H and Future Farmers of America (FFA) youth because—compared to other project species such as

The "egg toss" isn't just for family picnics. An organization—the World Egg Throwing Federation—actually exists to regulate games related to this old-fashioned pastime.

American Poultry Association Classifications

The APA further divides classes depending on each chicken's size and type.

Large-Breed Classifications
There are five classifications for large breed fowl: American, Asiatic, Mediterranean, Continental, and All Other Standard Breeds.
American Class: Buckeye, Chantecler, Delaware, Dominique, Holland, Java, Jersey Giant, Lamona, New Hampshire, Plymouth Rock, Rhode Island Red, Rhode Island White, and Wyandotte
Asiatic Class: Brahma, Cochin, and Langshan
English Class: Australorp, Cornish, Dorking, Orpington, Redcap, and Sussex
Mediterranean Class: Ancona, Andalusian, Catalana, Leghorn, Minorca, Sicilian Buttercup, and Spanish
Continental Class: Barnevelder, Campine, Crevecoeur, Faverolles, Hamburg, Houdan, La Fleche, Lakenvelder, Polish, and Welsummer
All Other Standard Breeds: Ameraucana, Araucana, Aseel, Cubalaya, Frizzle, Malay, Modern Game, Naked Neck, Old English Game, Phoenix, Shamo, Sultan, Sumatra, and Yokohama

Bantam Classifications
Bantams are divided into a different set of classifications, mostly dependent on the leg feathering (or lack thereof).
Modern Game Class: Modern Game
Game Class: American Game and Old English Game
Single Comb Clean Legged: Ancona, Andalusian, Australorp, Campine, Catalana, Delaware, Dorking, Dutch, Frizzle, Holland, Japanese, Java, Jersey Giant, Lakenvelder, Lamona, Leghorn, Minorca, Naked Neck, New Hampshire, Orpington, Phoenix, Plymouth Rock, Rhode Island Red, Spanish, and Sussex
Rosecomb Clean Legged: Ancona, Antwerp Belgian (or Belgian d'Anvers), Dominique, Dorking, Hamburg, Leghorn, Minorca, Redcap, Rhode Island Red, Rhode Island White, Rosecomb, Sebright, and Wyandotte
All Other Clean Legged: Ameraucana, Araucana, Buckeye, Chantecler, Cornish, Crevecoeur, Cubalaya, Houdan, La Fleche, Malay, Polish, Shamo, Sicilian Buttercup, Sumatra, and Yokohama.
Feather Legged: Booted, Brahma, Cochin, Belgian d'Uccle, Faverolles, Frizzle, Langshan, Silkie, and Sultan

cattle, sheep, and pigs—chickens are easy and inexpensive to buy, raise, and show. Youth poultry shows, whether 4-H, FFA, or nonsanctioned events, usually include classes for both purebred chickens and production birds and are judged by the modified Danish system, whereby each entry is awarded a blue, red, or white ribbon according to merit.

Your First Show

Let's assume you'd like to show your hen in the open poultry show at your county fair. If you plan to show a purebred chicken, take a long, hard look at your bird, comparing her to her breed's description in the *Standard of Perfection* (your library probably has a copy or can get one through interlibrary loan). No chicken is perfect, but you don't want to show a bird with obvious disqualifications, even at a county fair, where people routinely gain experience before tackling the big sanctioned shows.

Next, call or visit your county extension agent and request a premium book listing dates and times, the divisions and classes you can show in, what you can win, and any other particulars pertaining to the show. Study it and decide which classes are right for you. Check to see if your chicken needs health papers in order to show; if she does, obtain them in plenty of time. Fill in the entry form, write out a check for entry fees, and mail it to your county fair association. There, you're committed!

Judges handle each chicken to assess its merits, so begin taming your show bird right away. Start by teaching your chicken to stay calm and relaxed in a show-type confinement cage. At least a week before show day, practice removing her from the cage, head first, keeping her wings folded and close to her body. Hold her quietly for a time, then open her wings and examine various parts of her body. When you're finished, return her, head first, to her cage. Practice until

Make sure that your chicken is easily handled and well-behaved in close quarters before you enter it in any shows.

Make sure your chicken is ready for its close-up before you head out to a show. Once it's free of all dirt and debris, stay on top of your coop-cleaning duties to avoid too many last-minute touch-ups.

she remains cool and unruffled from start to finish.

A few days before the show, carefully evaluate your chicken's cleanliness. Birds with light-colored plumage nearly always need to be bathed, while dark or patterned chickens such as Dominiques and Barred Plymouth Rocks usually don't. A double sink works fine for bathing a few chickens. Start by filling one side with warm water laced with a little detergent or baby shampoo. Fill the other side of the sink with a warm-water rinse and ¼ cup of white vinegar (vinegar helps remove the soap).

Bring in your chicken and immerse her in the bath water, holding her wings against her sides so that she can't flap. Don't dunk her head under water. Soak her, gently moving her feathers so that the soap and water can penetrate clear to her skin. Spot-clean any dirty spots by rubbing a dab of detergent or shampoo into her feathers from base to tip, then gently scrub her legs using an old toothbrush.

When she's all cleaned up, transfer your chicken to the rinse water and rinse out as much soap as you can. After her rinse, let her soak for several minutes while you drain the wash water from the first side, wash out the sink, and refill it with warm water. Rinse her again in the fresh water, then wrap her in a terrycloth towel and blot away as much water as you can. Place her in a drying cage or use a hair dryer set on low heat to blow her dry (this works well to poof up the feathers of soft-feathered breeds, such as Cochins and Silkies, but don't use a hair

The Hen Says…

In most languages, hens' vocalizations are written onomatopoeically, meaning that they're spelled the way a clucking hen sounds. In English, for example, we say "cluck cluck." Here's what our "cluck cluck" looks like in some other languages.

Bulgarian: kudkudyak
Czech: kokodák
Dutch: tok tok
Finnish: kot-kot
French: cotcotcodet
German: tock tock
Greek: ko ko ko
Hebrew: chuck-chuck
Indonesian: petok petok
Italian: coccodè
Korean: gogode koko
Montenegrin: kokoda
Polish: baqka baqka
Romanian: cotcodac
Russian: ko-ko-ko
Spanish: klokloklokloo
Swedish: ock-ock
Turkish: gut gut gdak
Ukrainian: kud-kudak

The Rooster Says…

The same spelling rules apply to the vocalizations that roosters make. English-speaking roosters are said to say "cock-a-doodle-do." Here's what our "cock-a-doodle-do" looks like in some other languages.

Bulgarian: kukurigu
Czech: kykyryký
Dutch: kúkelekúú
Estonian: kikerikii
Finnish: kukkokiekuu
French: cocorico
German: kikeriki
Greek: kikiriki
Hebrew: kukuriku
Hungarian: kukurikúú
Indonesian: kukuruyuk
Italian: chichhirichì
Japanese: kokekokko
Lithuanian: kakariekūūūū
Montenegrin: kukuriku
Norwegian: kykkelikiii
Polish: kukuryku
Portuguese: cócórócócó
Romanian: cucurigu
Russian: kukuryku
Spanish: kikirikí
Swedish: kuckeliku
Turkish: üü ürüüü ürüüü
Ukrainian: kukuriku
Vietnamese: cuc-cuc-cu

dryer on close-feathered chickens, such as Old English Games or Cornishes).

After she's bathed, be extra-diligent about keeping her living quarters clean. The day before the show, spot-check for soiled feathers and massage a smidge of mayonnaise, olive oil, or vitamin E oil onto her comb, wattles, beak, and shanks, taking care not to spread oil onto her feathers.

Transport your chicken to the show in a sturdy but lightweight shipping crate lined with clean, dry straw or wood shavings on the bottom. Airline crates designed for small dogs work well. If you're taking more than one chicken, each needs its own container.

Arrive at the show grounds early, check in at the office, and find your chicken's assigned cage (in poultry-show termi-nology, this is called "cooping in"). Make sure that there are enough shavings on the floor of the cage (it never hurts to bring some extra from home, especially when showing feather-legged breeds) and fill your bird's water cup. Don't do anything to your bird's cage to distinguish your chickens from other people's birds, such as putting up signs or displays.

Take your chicken out of her carrier and look her over carefully before putting her in her cage. Check her eyes, nostrils, shanks, and toenails for dirt, and then stroke her feathers with a clean cotton or silk cloth to give her an extra shine. Pop your bird in her cage and relax! Take a break, admire the competition, and talk chickens with fellow exhibitors until the judging starts.

When judging begins, watch from a distance. Never interrupt the judge while he's working. After he's finished, you'll see a note explaining your placing or a ribbon attached to your chicken's cage.

Each show is a learning experience, and showing is fun. If you've placed well, congratulations! If not, come back next year and try again.

If all goes well, you'll have a champion on your hands by the end of the show day. If not, keep trying!

Chickens and Guinness World Records

(as of October 15, 2010)

- The largest chicken in the Guinness record book is 23-pound, 3-ounce "Big Snow," who was owned by Ronald Alldridge of Deuchar, Queensland, Australia. Big Snow died on September 6, 1992 of natural causes.

- Guinness claims that the oldest living chicken was a 14-ounce, 14-year-old (at the time of entry) hen named Matilda (1990–2006), owned by Keith and Donna Barton of Bessemer, Alabama. However, a Chinese farmer named Chen Yubin of Zhengzhou, Henan Province, recently filed documents authenticating the age of his 20-year-old hen.

- In 1945, a white rooster named Mike earned a place in the record book as the longest-living chicken without a head. Mike lived for eighteen months after being decapitated for Sunday dinner.

- The world's largest egg (currently pending acceptance by Guinness World Records) was laid by an everyday hen belonging to Zhang Yinde of Heilongjiang Province, China. The egg is 2.5 inches wide by 3.6 inches long, weighing 7.09 ounces.

- In March of 2009, Zachery George of West Virginia broke the old Guinness record for egg-holding by holding twenty-four eggs in his right hand.

- The world record for the number of people participating in an egg toss was set on July 4, 2008, in Skagway, Alaska, with 1,422 participants.

- On November 12, 1978, in Jewett, Texas, Johnny Dell Foley successfully tossed a raw egg 323 feet and 2 inches to Keith Thomas, who caught it unbroken. Their record was undefeated until at least 1999. (Since 2000, the feat is no longer listed in the book.)

- On October 11, 2010, the Turkish Egg Producers Association celebrated World Egg Day by frying a world-record omelet that weighed 13,227.7 pounds using 114 gallons of oil and 110,000 eggs.

- On August 19, 2010, Kentucky Fried Chicken treated the citizens of Louisville, Kentucky, to the world's largest serving of fried chicken, weighing in at 2,493.35 pounds, thus breaking a record set one month earlier by volunteer cooks at Canoefest in Brookville, Indiana, who fried 1,645 pounds of donated chicken served up in a donated canoe.

Appendix I: Chicken Stories

To some people, chickens are simply meat or egg makers. To others, intelligent, affectionate chickens are the best darned pets on earth. These stories were written by members of the "housechickens" email group. Read them, then tame a special chicken or two of your own. Join housechickens (see Resources) and share your stories. Perhaps they'll appear in a later edition of this book.

Piggy Hero

About three and one-half years ago, I was living on the family farm in Brooks, Maine, and attended a local country fair, where I fell in love with a fluffy black Silkie rooster who was sitting in a ten-year-old girl's lap, complacently allowing himself to be petted. I went back and forth from that poultry exhibit (always my favorite part of any country fair, followed closely by prize-winning veggies) and finally went home with that roo-boy in a cardboard box.

That roo was Piggy (originally named Suleyman) and since I had only the one, and no chicken house set up, and because I like all manner of living things in my house, he became a very pampered and beloved house pet, and in many ways my very best friend. Piggy has never been constrained in any way, except to the house (though, of course, he does have supervised yard forays) and so has developed his own routines. When he's not asleep, Piggy socializes. Wherever the people are is where he will be, either sitting behind them on the back of the couch watching videos, or nestled on the bed beside a lazy reader, or pattering from room to room as errands are tended to.

But at night, Piggy makes his way to his bed, which is the back of the couch (eternally towel-draped), and, generally speaking, he will then sleep through all manner of activity and noise. Now and then, he'll arise with a second wind if the lights are turned on or if he hears activity in a distant room and finds it too quiet where he is, but generally he sticks to his routine. In the morning, he hops up, crows, gets himself a bite to eat, and seeks out a bed to jump into, where he will greet any sign of movement by gently nibbling noses or tugging on hair in an attempt to waken a sleeper for a petting session. If he is unsuccessful in his attempts, as he often is, he will settle down to nap quietly until his companions finally arise. Often, I have walked by my significant other's room to see a heap of bedclothes, with nothing emerging but Piggy's rotund rump and a rather limp hand petting said rump.

One night, Mark was away, and our housemate Paul and I were the only humans in the house. Piggy had gone to bed as usual several hours before, and Paul had retired to his upstairs bedroom. I was in my room at the back of the house, picking sluggishly at a couple watercolor commissions. Around 11:30, I passed Piggy on the way to the kitchen to put on the tea kettle, but Piggy never woke up, although I briefly rubbed his back in passing, as always.

It can't have been two minutes later that I heard Piggy hit the floor with a thud. This caught my attention because it was late and there was no conversation for him to be seeking out, no lights, nothing to wake him. I called his name. Suddenly he rushed into my room and leapt onto the bed, clearly agitated and making distress noises. Thinking vaguely that perhaps an animal had gotten in and frightened him, I ran through the living room to find the kitchen half engulfed in flames, visible only as a large red glow through thick black smoke. I tried to turn on the light and thought the bulb must have blown. (I later discovered that it was still working, but the smoke was so thick that I could not tell if it was on or off.)

I screamed for Paul and fumbled with the outside door until it opened, got the outside light on, and tore open the door of the bulkhead to the basement, where the garden hose had recently been stored. As Paul dialed 9-1-1, I turned the hose on full force, aiming it at the door of the kitchen as smoke pumped out like a huge black caterpillar.

By the time the fire trucks arrived, the walls and ceiling were still smoldering, but I had gotten the flames out some minutes before. We were hustled from the house and the fire axes went to work. At one point, I sneaked in and checked on Piggy, and he was standing alertly on the bed where I last saw him, so I left him there and went back outside while the firemen finished up.

They said that surely, after another minute, we would have lost the whole house and—they did not need to say it—possibly our lives, as well. They asked what had happened and I told them all, including Piggy's role. As they were winding down, I was standing in the smoldering remains of my kitchen, talking to two firemen in full gear, when one of them cried, "There's the chicken!" Piggy, hearing my voice, had taken that for an all-clear, and left the refuge of my bed to amble out, rather timidly by his own standards, to be introduced.

The next evening, Piggy, who when he is not being a hero is just a regular stay-at-home kind of guy, ate a prodigious quantity of his favorite food—a peanut butter, cream cheese, and olive sandwich (you read right)—wiped his beak repeatedly on my leg, and snuggled down beside me for a nap. I intend to propose that a monument be erected in Piggy's likeness on the fire-station lawn!

—Cindy Ryan

Cindy Ryan is a School of Visual Arts (NYC)-trained artist and children's writer.

Punky Rainbow

Spring 2001 to March 26, 2003
Member of the Kroll Family

Love with Wings for $1.25

There was something about her.

When she came home she followed me everywhere.

Punky Rainbow became a house chicken for a while.

She would sing and follow me around the house.

Punky kept her little wings outstretched like

preparing for take off as she and our family

played tag up and down the hallway.

We trusted each other completely.

She was so gentle.

I could pick her up without a single flap.

Punky Rainbow would preen my eye lashes.

Many city folk met her at our pet store;

Punky was the first living chicken they had ever seen.

When she moved to the farm to have babies,

Punky and I still sat and talked.

She seemed to understand what I said,

and chirped and purred back to me.

She went with us camping, to the beach,

sledding, and to the pet store.

"Just this side of Heaven is a place called the Rainbow Bridge where
Punky Rainbow is playing. We'll cross the Rainbow Bridge together."

— *Jennifer Kroll, Fluff 'N Strut Silkies*

A Tribute to Eggnus (2003–2004)

Eggnus came into our lives on April 6, 2003. I had placed an order for two chicks at a feed store and when we picked them up, I saw a box labeled "White Rocks." Inside it were the cutest yellow chicks with a touch of peach on their chests! One chick looked up at me, almost pleading for me to pick her, too. I couldn't resist; I had found my Eggnus.

I had a Silkie hen named Emma, who was raising two chicks of her own, and Emma adopted Eggnus, who immediately cuddled under Emma and joined the other chicks. All through Eggnus' life, she remained close to her foster mom.

However, Eggnus quickly outgrew her mom and the other large-breed chicks. She would trail behind the others, and when she'd run, we called it the "wiggle waddle." She would hop back and forth sideways, and she had trouble stopping when she started running; once, she even knocked down a rooster who was standing in her path!

When she was four months old, Eggnus went lame. I took her to a vet. Eggnus was the talk of the waiting room—this big white chicken in a dog crate. She made friends with everyone at the vet's office, including the vet's assistant, who told me Eggnus was a Cornish/Rock cross.

The vet thought Eggnus had injured her back jumping down out of the coop because she was obese. At four months old, she weighed over nine pounds! She was given vitamin shots and medicine, and I put her on a diet.

After that, Eggnus would go outside in the mornings with the flock, then she'd come in the house, where she had her own personal fan and lived in a plastic crate in the hallway. She'd flash those big eyes at you and tilt her head sideways. She loved to be petted and greeted everyone who passed by.

Eggnus never regained her ability to walk any long distances. She would take a couple of steps and then lie down. But that never stopped her from trying. Eggnus was determined to walk.

Eggnus died January 3, 2004, in her sleep. I knew from the research I did on her breed that she wouldn't live to be very old. However, knowing this didn't make her passing any easier. We buried her beside a peony bush she had loved to lie beneath.

I knew Eggnus was special the first time I saw her—I just didn't know how special she really was. She always had a cheerful attitude and the sweetest personality. I will never know why a Cornish/Rock cross was in a box with White Rock chicks, but if she hadn't been, I would never have known my Eggnus. I have no regrets. If I had known what I know now, I would do it all over again.

A gentle giant of a chicken with a heart of gold: Eggnus, I will love and remember you always.

—*Patty Mousty*

Appendix II: Pick of the Really Cool Chicks

f you're like me, you like chickens with pizzazz—something that catches people's eyes. There are dozens of breeds that fit that description, so it was hard to choose, but these are my picks of the really cool chicks.

Dorkings

If you want to raise a piece of living history but you need productive birds as well, Dorkings are your chickens! These short-legged, five-toed fowl were carried to Great Britain by Julius Caesar's troops, making them one of the world's oldest recognizable breeds. Dorkings were further developed in England and among the first chickens imported to the American colonies. They are named for the market town of Dorking in the south of England.

Dorkings are among the breeds listed in the "Threatened" category on the American Livestock Breeds Conservancy's Conservation Priority List. They come in four recognized varieties: White, Silver Gray, Colored, and Rose-Combed Cuckoo. They have rectangular, meaty bodies set on very short legs. Dorkings are good layers of tasty, creamy, white or palely tinted eggs—even through the cold winter months—and they're also noted for their luscious meat. Hens are notoriously broody and are excellent moms. Gentle, unflappable, dual-purpose chickens that thrive in confinement or as free-range fowl, Dorkings are wonderful pets and great city chickens.

Dutch

Diminutive Dutch bantams are among the smallest of chickens. Their breed standard calls for 21-ounce roosters and 19-ounce hens. They are true bantams; there is no standard-size version of this fine old breed.

The ancestors of today's Dutch bantams were probably introduced to the Netherlands by Dutch East India Company sailors returning from Asia, though their exact origin is unclear. Over several hundred years, Dutch breeders selected for small, attractive fowl that laid a lot of eggs. They are still outstanding layers of petite, creamy or light-brown eggs, averaging 150 to 160 tasty eggs per year.

Dutch come in an array of colors. They're flighty but friendly and are noted for their ability to fly. Hens brood their own eggs and are good, protective moms. Dutch make fine pets and city chickens. If you crave chickens but don't have a lot of space to keep them, Dutch bantams could be the answer to your prayers.

Jersey Giants

If you want useful chickens and you want them supersized, Jersey Giants are the birds for you. Developed between 1870 and 1890 by the Black brothers of Jobstown, New Jersey, Jersey Giants head the list of the world's largest breeds.

A typical Jersey Giant hen tips the scale at 10 pounds; roosters grow to 13 pounds or better. Hens are good layers of extra-large, light cream-colored to dark brown eggs that they'll often happily hatch, given a chance to go broody.

Jersey Giants are also excellent eating fowl, though they're relatively slow to fill out and take eight to nine months to reach roasting size.

Giants are calm, friendly, dual-purpose chickens that do equally well as pets, city chickens, or resourceful, free-range birds in the country. They come in three recognized colors (Black, Blue, and White) as well as Splash (splash-colored chickens range from predominantly white offset with a few splotches of blue, to nearly solid blue with white accents). They are listed in the American Livestock Breeds Conservancy's "Watch" category, meaning that they're uncommon but not critically endangered.

Naked Necks

Naked Necks—you'll love them or you'll hate them. With their bare necks and sparsely-feathered bodies, they aren't very pretty, but they're unique and very useful birds.

Also known as Transylvanian Naked Necks or Turkens, the breed originated in Central Europe, though German breeders in the nineteenth century developed their modern form. They're popular in Germany and France but relatively rare in North America. Naked Necks are listed in the American Livestock Breeds Conservancy's "Study" category.

In addition to their feather-denuded necks, these chickens have less than normal feathering on their breasts, under their wings, and around their vents. The reason: their relative lack of feathers makes them easy to pick. They are excellent table birds and also respectable layers of large brown eggs, making them uncommonly useful dual-purpose fowl. Hens are also broody and are extra-good moms.

Naked Necks are both heat and cold hardy, though their bare necks are subject to sunburn. They are excellent foragers and are active, but they are also calm and friendly. Naked Necks make fine conversation-starting pets and city chickens as well as productive free-ranging country birds.

Phoenix

Phoenix and Yokohama

Looking for a gentle, elegant, ornamental chicken with beauty and novelty to spare? Look no further: Phoenix and Yokohama chickens fit the bill. Both breeds descend from long-tailed Japanese fowl imported to Germany prior to 1900. Their honorable distant ancestor is the Onagadori chicken of Japan, a breed noted for roosters' tails up to 27 feet long!

Though separate breeds, Phoenix and Yokohama chickens are much alike. They are small, regal birds that weigh in the neighborhood of 4 to 5 pounds. If provided with extra protein in their diets, they are fair layers of smallish, white to lightly tinted eggs. Phoenix have single combs and come in Silver, Golden, and Black-Breasted Red; Yokohamas have walnut combs and come in White and Red-Shouldered White. Roosters have long hackle feathers and grow 3- to 4-foot-long tails if kept inside and provided with perches so their tails don't drag. They are alert but calm and fairly quiet breeds, making them suited for life in the suburbs. Both are listed on the ALBC's Conservation Priority List, Yokohamas on the "Critical" list and Phoenix in the "Threatened" category.

Yokohama

Polish

Polish are the ultimate in crested chickens. Some Polish varieties are bearded, as well. But they aren't merely odd-looking ornamentals; some (though not all) Polish hens are fine layers of large white eggs, and they're good pet chickens, too.

Polish chickens did not originate in Poland. Their history is quite obscure, but many poultry historians believe that they were carried from Spain to Holland when Spanish settlers occupied the lowlands. Fowls similar to today's Polish chickens figure in works by Dutch and Italian artists painted in the sixteenth through eighteenth centuries. They were known in England during the 1700s and were well established as layers in America by the mid-1800s.

Slim and racy-looking, Polish are generally calm and friendly fowl, but their crazy bouffant "hairdos" obscure their vision, so these birds are somewhat easily startled. Frozen crest feathers can be a problem, so they aren't suited for cold, wet conditions. Polish come in an array of colors and have V-shaped combs. They are fairly uncommon and currently listed in the American Livestock Breeds Conservancy's "Watch" classification.

Russian Orloffs

Russian Orloffs, once recognized simply as "Russians" by the American Poultry Association, are large, fierce-looking, bearded, and muffed chickens with short, hooked beaks and heavy eyelids. They descend from fowl taken from Persia (Iran) to Russia prior to the seventeenth century, where they were originally called Chlianskaia chickens. They are presently named in honor of the former commander-in-chief of the Russian navy, Count Alexei Orloff, who promoted them during the nineteenth century.

Their looks are deceiving; these are calm, good-natured birds that fare well as free-range fowl or held in winter confinement. They have tiny walnut combs, a characteristic that helps make them exceptionally cold hardy. Originally developed as meat birds, Orloff hens lay a respectable number of medium-size pale-brown eggs right through the cold winter months.

Though other colors occur in Britain and Europe, Russian Orloffs come in a single color, Spangled (speckled black and white), in America. Though they're gaining in popularity, Russian Orloffs are among the rarest breeds classified by the American Livestock Breeds Conservancy as "Critical."

Seramas

Seramas are the world's smallest chickens and are rare in the United States at the time of this writing. They may also be the world's best pet chickens, so if you want a house fowl, a Serama could be your bird.

Seramas originated in Kelantan, Malaysia, where they are still common household pets. They are graded according to size, with roosters weighing a mere 12.8 to 19.2 ounces and hens between 11.3 and 16.8 ounces. Seramas are tiny; a pair or trio fits nicely in a 24- by 18-inch cage. Because they hail from the hot, steamy jungles of Malaysia, Seramas require extra heat in their cages (a 40-watt incandescent bulb does the trick). Having been developed as pets, Seramas are calm, ultra-friendly little birds that crave human attention. A bonus: hens are broody and make great moms.

Seramas come in every color combination known to chickens. They are incredibly chesty and short-backed, and both sexes stand up in a nearly vertical pose with their tails up high and their wings pointed down at their sides. Along with being teensy, they are very cute.

Silkies

Silkies are the quintessential pet chickens. Sweet and docile, they beg to be held. They're also noted for their overwhelming broodiness, making them peerless living incubators for hatching eggs of many species, from tiny quail eggs to those of geese.

Fluffy five-toed Silkies are unique in many ways, being the only chickens in America with dark-blue flesh and bones (their Chinese name, wu gu ji, means "crow-boned chicken"). They are bantam-size fowl with feathered legs, turquoise blue earlobes, dark wattles, and petite, walnut-type combs. Silkie feathers resemble down, which lacks barbicels (the minute hooks that hold feathers together in the shape we usually equate with chicken feathers). This gives Silkies their distinctive look and feel. Silkies come in an array of colors and in bearded and non-bearded varieties. They also lay a respectable number of small but tasty cream-colored to light-brown eggs.

Silkies are an ancient Oriental breed. Marco Polo wrote about "fur-covered" chickens in the thirteenth century, and Italian author Ulisse Aldrovandi spoke of "wool-bearing chickens" in 1599. The American Poultry Association recognized Silkies in their first *Standard of Perfection*, printed in 1874. Cute, sweet, and oh-so-unique, Silkies have been around for a very long time.

Sultans

The weird and wonderful Sultan's ancestors hail from Turkey, where the breed is called the Serai Taook or "Sultan's Fowl." They arrived in England in 1854 and, from there, came to the United States in 1867. Sultans were included in the first volume of the American Poultry Association's *Standard of Perfection*, published in 1874.

According to the American Livestock Breeds Conservancy (Sultans are included on its Conservation Priority List in the "Critical" classification), these fowl have more distinguishing characteristics than any other breed, including a V-shaped comb, a crest, a beard, muffs, five toes, feathered shanks and toes, large nostrils, wings carried low, and vulture hocks. Roosters tip the scale at 6 pounds; hens are lighter at about 4 pounds. Only one color is standard: white with slate blue shakes and toes. Hens lay a fair number of large white eggs and are nonbroody. Once noted as delectable table birds, they have large breasts and delicate, white flesh.

Sultans are unusually calm and friendly chickens, content to be housed in confinement. If you're seeking an unusual ornamental or pet chicken, Sultans are a first-rate choice.

Appendix III: Layers at a Glance

This chart includes breeds that lay at least three eggs per week in top production. Keep in mind, however, that actual productivity varies by season and individual hen.

Layers at a Glance

	Origin/Admitted to APA /Rarity	Type	Weight (Hen)	Colors/Comb Skin Color/ Uncommon Characteristics
Ameraucana	South America/1984/ purebreds are uncommon	Layer	5.5 lbs.	Black, Blue, Blue Wheaten, Brown Red, Buff, Silver, Wheaten, White/pea/ white/muffed, bearded
Ancona	Italy, mid-19th c./1898/ALBC listed in Watch category	Layer	4.5 lbs. (bantam 1.5 lbs.)	Black mottled/single and rose/yellow/none
Andalusian	Spain, mid-19th c./1874/ALBC listed in Threatened category	Layer	5.5 lbs. (bantam 1.75 lbs.)	Blue/single/yellow/none
Australorp	Australia/1929/common	Dual purpose	6.5 lbs. (bantam 2 lbs.)	Black/single/white/none
Barnevelder	Holland, 19th c./2001/ uncommon	Dual purpose	6-7 lbs. (bantam 2.25 lbs.)	Double Laced, Blue Laced, White, Black/single/ yellow/none
Black Sex Link	United States, 20th c./not recognized/common	Layer	About 5 lbs.	Black with golden hackle/single/ yellow/none
Brahma	Probably China or India/1874/ ALBC listed in Watch category	Dual purpose	9 lbs. (bantam 2.6 lbs.)	Light, Dark, Buff/pea/ yellow/feather-footed
Buckeye	United States, late 19th c./1904/ ALBC listed in Critical category	Dual purpose	6.5 lbs.	Dark nut brown/pea/ yellow/none
Buttercup	Sicily, mid-19th c./1918/ ALBC listed in Threatened category	Layer	5 lbs.	Golden/buttercup/ yellow/none
Campine	Belgium, 19th c./1914/ ALBC listed in Critical category	Layer	5 lbs.	Golden, Silver/large single/white/none
Catalana	Spain, late 19th c./1949/ ALBC listed in Watch category	Layer	6 lbs.	Buff/single/yellow/none
Chantecler	Canada, early 20th c./1949/ ALBC listed in Critical category	Dual purpose	6.5 lbs.	White, Partridge/ cushion/yellow/none
Cochin	China, very old breed/1874/ ALBC listed in Watch category	Dual purpose	8.5 lbs. (bantam, a.k.a. Pekin, 1.3 lbs.)	Buff, Partridge, Black, White/small single/ yellow/feather-footed

Egg Color/ Size/ Productivity*	Behavior	Hardiness	Adaptable to Confinement?	Broody?	Good City Chicken?
Shades of blue and blue-green/average/ very good	Calm	Cold hardy	Yes	Occasionally	Yes
White or tinted/small/ excellent	Nervous, active flyer	Rose combed variety especially cold hardy	Sometimes	No	No
White/large/good	Flighty, active flyer, noisy	Heat hardy	Sometimes	No	No
Brown/medium/ very good	Quiet, calm	Cold hardy	Yes	Yes	Yes
Vary dark reddish-brown/ medium-large/good	Calm	Reasonably heat and cold hardy	Yes	Sometimes	Yes
Brown/large/excellent	Calm	Reasonably heat and cold hardy	Yes	No	Yes
Brown/medium-large/good	Calm	Heat and cold hardy	Yes	Yes	Yes
Brown/ medium-large/ good	Calm, friendly	Very cold hardy	Yes	Yes	Yes
White/medium/good	Flighty, wild	Heat hardy	Sometimes	No	No
White/medium-large/good	Alert, semi-flighty, active flyer	Cold hardy (combs prone to frostbite)	Sometimes	No	No
White or light tinted/ medium-large/ very good	Active, semi-flighty	Heat hardy	Sometimes	No	No
Brown/medium-large to large/good (generally lays year round)	Calm, friendly	Very cold hardy	Yes	Yes	Yes
Tinted to light brown/ medium/good	Calm, friendly	Heat and cold hardy	Yes	Yes	Yes

*Good = about 3 eggs per week, very good = about 4 eggs per week, excellent = about 5 or more eggs per week.

Crevecoeur	France, pre-18th c./1874/ ALBC listed in Critical category	Dual purpose	6.5 lbs.	Black/V-shaped/white/ crested, bearded, and muffed
Delaware	United States/1952/ ALBC listed in Threatened category	Dual purpose	6.5 lbs. (bantam 1.8 lbs.)	White/single/ yellow/none
Dominique	United States, early 18th c./1874/ ALBC listed in Watch category	Dual purpose	5 lbs. (bantam 1.5 lbs.)	Barred/rose/yellow/none
Dorking	Ancient Rome/1874/ ALBC listed in Threatened category	Dual purpose	6.5 lbs. (bantam 1 lbs.)	White, Silver Gray/rose and single/white/very short legs and 5 toes
Faverolles	France, early 19th c./1914/ ALBC listed in Threatened category	Dual purpose	7.5 lbs./bantam 2.2 lbs.	White, Salmon/small single/white/bearded, muffed, feather footed. 5 toes
Fayoumi	Ancient Egypt/not recognized/un-common	Layer	3.5 lbs.	Gold Penciled, Silver Penciled/single/white/none
Hamburg	Holland before 1700/1874/ ALBC listed in Watch category	Layer	4 lbs. (bantam 1.5 lbs.)	Gold Spangled, Silver Spangled, Gold Penciled, Silver Penciled. Black, White/rose/white/none
Holland	United States, 20th c./1949/ ALBC listed in Critical category	Dual purpose	6.5 lbs.	White, Barred/single/ yellow/none
Java	United States, 19th c.)/1874/ ALBC listed in Threatened category	Dual purpose	7.5 lbs.	Black, Mottled/single/ yellow/none
Jersey Giant	United States, late 19th c./1922 / ALBC listed in Watch category	Dual purpose	10 lbs.	Black, White/small single/ yellow/huge size
La Fleche	France prior to mid 17th c./1874/ ALBC listed in Watch category	Dual purpose	5.5 lbs.	Cuckoo, Blue, Black, White/ V-shaped/white/none
Lakenvelder	Germany, early 19th c./1939/ ALBC listed in Threatened category	Layer	4 lbs.	Black and white pat-terned/single/white/none
Langshan	China (to England in mid-19th c.)/1883/ ALBC listed in Threat-ened category	Dual purpose	7.5 lbs.	Black, Blue, White/single/ grayish white/feather-footed
Leghorn	Italy (further developed in England and United States in 19th c.)/1874/common	Layer	4.5 lbs.	White, Black, Light brown, Dark brown, Buff, Silver, Columbian, Red, Black-tailed Red/large single and rose/ yellow/none
Marans	France, early 20th c./not recog-nized/uncommon	Layer	7 lbs. (bantam 2.2 lbs.)	Dark, Cuckoo/single/ white/some strains are feather-footed

White/medium-large/good	Active, alert	Not suited for wet conditions	Yes	No	Possibly
Brown/large/very good	Calm	Heat and cold hardy	Yes	Sometimes	Yes
Brown/medium-large/good	Generally calm	Cold hardy	Yes	Yes	Yes
Light tint/medium-large/good	Calm	Cold hardy	Yes	Yes	Yes
Light tint/medium-large/very good	Calm	Cold hardy but not good in wet conditions	Yes	Sometimes	Yes
Light tint/medium-small/very good	Flighty, somewhat wild	Heat hardy	No	No	No
White/medium /very good	Flighty, active flyer	Cold hardy	Rarely	No	No
White/medium-large/good	Calm, friendly	Cold hardy	Sometimes	Yes	Yes
Brown/medium-large/good	Calm	Cold hardy	Yes	Yes	Yes
Brown/medium-large to large/good	Calm, friendly	Very cold hardy	Sometimes	Yes	Yes
Tinted/large/good	Flighty, active	Cold hardy	No	Yes	Yes
White or tinted/medium to medium-large/good	Flighty, active flyer	Reasonably cold hardy	No	Yes	Sometimes
Brown/medium-large/good	Active	Reasonably cold hardy	Sometimes	Yes	Sometimes
White/large to extra-large/excellent	Active, extra flighty, active flyer, noisy	Heat hardy	Sometimes	No	No
Dark brown/medium-large to large/good	Varies by strain	Reasonably cold and heat hardy	Yes	Yes	Sometimes

Minorca	Spain, prior to 18th c./1888/ ALBC listed in Watch category	Layer	7 lbs. (bantam 1.75 lbs.)	Buff, White, Black/single and rose/white/none
Naked Neck (Turken)	Central Europe prior to 19th c./1965/ ALBC listed in Study category	Dual purpose	6 lbs. (bantam 1.5 lbs.)	Black, White, Red/single/yellow/no feathers on neck and fewer feathers on body
New Hampshire	United States, early 20th c./1935/ ALBC listed in Watch category	Dual purpose	6.5 lbs. (bantam 1.9 lbs.)	Light reddish brown/large single/yellow/none
(Russian) Orloff	18th c. Russia from Persian stock/not currently recognized/ ALBC listed in Critical category	Dual purpose	6 lbs.	Spangled, Red, White/walnut/yellow/bearded and muffed
Orpington	England, 19th c./1902/ ALBC listed in Recovering category	Dual purpose	8 lbs. (bantam 2.2 lbs.)	Buff, Blue, Black, White/single/white/none
Plymouth Rock	United States, 19th c./1874/ ALBC listed in Recovering category	Dual purpose	7.5 lbs. (bantam 2.2 lbs.)	Barred, Partridge, Columbian, Blue, Silver Penciled, Buff, White/single/yellow/none
Polish	Europe prior to 16th c./1874/ ALBC listed in Watch category	Layer or ornamental	4.5 lbs.	Black, White, Golden, Silver, Buff Laced/small V-shaped/white/crested, some are bearded
Redcap	England, early 17th c./1888/ ALBC listed in Critical category	Layer	6 lbs.	Red and black patterned/very large rose with spikes/white/none
Red Sex Link	United States, 20th c./not recognized/common	Layer	About 5 lbs.	Red with flecks of white/single/yellow/none
Rhode Island Red	United States, 19th c./1904/ common	Dual purpose	6.5 lbs. (bantam 2.2 lbs.)	Reddish brown/large single and rose/yellow/none
Spanish	Spain before 19th c./1874/ ALBC listed in Critical category	Layer	6 lbs.	White-faced Black/large single/gray/none
Sussex	England, 19th c./1914/ALBC listed in Recovering category	Dual purpose	7 lbs. (bantam 2.2 lbs.)	Speckled, Red, Light/single/white/none
Welsumer	Holland, 20th c./2001/ uncommon	Dual purpose	6 lbs. (bantam 2 lbs.)	Red Partridge/small single/yellow/none
Wyandotte	United States, 19th c./1883/ ALBC listed in Recovering category	Dual purpose	6.5 lbs. (bantam 2.2 lbs.)	Silver Laced, Gold Laced, Silver Penciled, Columbian, Partridge, Buff, White/rose/yellow/none

White/large/very good	Active, flighty	Heat hardy	Sometimes	No	No
Brown/medium-large to large/good	Active but calm	Heat and cold hardy. Neck can sunburn.	Yes	Yes	Yes
Brown/medium-large to large/excellent	Calm	Heat and cold hardy (combs subject to frostbite)	Yes	Yes	Yes
Brown/medium/good	Calm	Cold hardy	Rarely	Yes	Yes
Brown/medium-large to large/good	Calm, friendly	Cold hardy	Yes	Yes	Yes
Brown/large/good	Calm, friendly	Cold hardy	Yes	Yes	Yes
White/medium/varies widely from poor to very good	Usually calm; some strains said to be flighty	Not suited to cold, wet conditions	Yes	No	Usually
White/medium/good	Active	Reasonably cold and heat hardy	Yes	No	Yes
Brown/large/excellent	Calm	Reasonably cold and heat hardy	Yes	No	Yes
Brown/large/unusually excellent	Active but calm	Heat and cold hardy (large single combs prone to frostbite)	Yes	Occasionally	Yes
White/large/good	Calm to flighty, noisy	Heat hardy. Large combs prone to frostbite	Sometimes	No	No
Light brown/medium-large/excellent	Active but calm	Cold hardy	Yes	Yes	Yes
Dark brown/large/good	Calm	Cold hardy	Yes	Varies by strain	Yes
Brown/large/good	Calm	Very cold hardy	Yes	Sometimes	Yes

Glossary

Albumen—egg white

Alektorophobia—(also spelled Alektrophobia) the fear of chickens

American Bantam Association (ABA) —an organization devoted to the standardization and promotion of bantam chickens and ducks

American Poultry Association (APA)— an organization devoted to the standardization and promotion of large size chickens and ducks

Avian—having to do with birds; avian medicine, avian art, etc.

Bantam—a miniature chicken weighing 1 to 3 pounds and 1/4 to 1/5 the size of large breed chickens; some are scaled-down versions of large breeds but a few bantam breeds have no full-size counterparts; slang: banty (plural: banties)

Barny—a mixed-breed chicken; also called a barnyard chicken or, if bantam size, a barnyard banty

Beak—the protruding mouth parts of a bird

Beard—the tuft of feathers under the beaks of muffed breeds such as the Ameraucana, Belgian d'Uccle, and Houdan

Bedding—the thick layer of cushiony, absorbent material used to line nest boxes and poultry building floors; also called litter

Biddy—affectionate slang for hen

Billing out—the act of chickens using their beaks to scoop feed out of a feeder and onto the floor; filling feeders only partly discourages this wasteful habit

Bleaching—the fading of color from a aging, yellow-skinned laying hen's beak, shanks, and vent

Bloom—the protective coating on an unwashed egg

Blowout—vent damage caused by laying a huge egg

Booted—having feathers on shanks and toes

Breaking up—the act of convincing a broody hen that she doesn't want to set

Breed—a group of birds or animals having approximately the same background, body shape, and features; breeds "breed true" when an offspring from two individuals of the same breed resemble her parents

Breeders—birds from which fertile eggs are collected for artificial incubation purposes; also, people who breed chickens

Broiler—a tender nine- to sixteen-week-old meat chicken generally weighing 2.5–3.5 pounds; also called a fryer

Brood—(verb) to keep chicks warm inside a heated enclosure called a brooder or brooder box, or under a hen; (noun) the young of a chicken

Broody hen—one who, through hormonal changes, stops laying and elects to hatch eggs or care for baby chicks; also called a broody

Candle—to evaluate the contents of an egg by holding it up to a strong light source

Candler—the strong light source used to candle eggs

Cannibalism—the habit stressed chickens have of pecking other chickens or themselves until they draw blood; a bloodied chicken often is killed by her peers

Cape—the narrow feathers between a chicken's neck and back

Capon—a castrated male chicken

Cecum—a pouch-like internal organ located where small and large intestines connect

Chalazae—the white, stringy cords on opposite ends of an egg yolk that center it within the egg

Check—a lightly cracked egg with its inner membranes still intact

Chook—originally Australian slang; a common synonym for chicken

City Statutes—Laws defining what can be done within the bounds of a specific city limits, including whether or not homeowners can keep chickens, and if so, how they must be housed and maintained

Class—a group of chickens developed in a certain locale (e.g., American, Asiatic, Continental classes); also a group of chickens being judged together at a show

Clean-legged—having no feathers on the shanks

Cloaca—the chamber inside a chicken's vent where her digestive, excretory, and reproductive tracts meet

Cluck—a hen vocalization; also an affection word for hen

Clutch—a batch of eggs hatched in an incubator or under a hen; all the eggs laid by a hen on consecutive days before she skips a day to begin a new laying cycle

Coccidiostat—a drug used to prevent the common protozoal infection, coccideosis; often added to commercial chick starter rations

Cock—a one-year-old or older male chicken; also called a rooster

Cockerel—a male chicken less than one year old

Comb—the fleshy, red appendage atop a chicken's head

Condition—a chicken's state of health, weight, and cleanliness

Conformation—an animal or bird's body structure

Coop—a building where chickens reside

Crest—the wild topknot of feathers adorning the heads of crested breeds such as Polish, Crevecoeur, Houdon, and Silkie chickens; also called a topknot

Crop—the expandable pouch of tissue in a chicken's lower esophagus where food is temporarily stored before digestion begins

Crossbred—also called a crossbred; a chicken with parents of two different breeds

Cull—to eliminate a bird from the flock

Debeak—to burn or snip off the tip of a chicken's upper beak to prevent cannibalism

Dressed—prepared for cooking: decapitated, picked, eviscerated

Dual-purpose breed—one who doesn't lay as many eggs as laying breeds nor mature as fast or as large as hybrid meat breeds but who lays more eggs than meat breeds and makes more meat than layers; an all-rounder

Dub—to trim a rooster's comb and snip off his wattles. Sometimes done to prevent frostbite or remove frostbitten tissue; Old English Game cocks must be dubbed in order to be shown

Dust bathing—the act of a chicken wallowing in dirt to clean its feathers and discourage external parasites

Earlobes—the patches of fleshy, bare skin below a chicken's external auditory meatuses

Egg tooth—a minuscule, sharp projection on a hatching chick's beak used to peck holes in her shell

Embryo—an unhatched, developing chick

Embryology—a branch of science devoted to the study of embryonic development

Exhibition breeds—fancy chickens raised for show instead of production

Feather-legged—also called feather-footed; having feathered shanks, feet, or both

Fertility—the state of being fertile

Finish—the amount of fat beneath the skin of a broiler or roaster chicken

Flock—a group of chickens

Free choice—when feed, water, or any other substance is available at all times

Free-range chickens—uncaged fowl allowed to forage wherever they choose

Frizzle—feathers that curl instead of laying flat; also a specific breed of chicken having frizzled feathers

Fryer—a tender nine- to sixteen-week-old meat chicken generally weighing 2.5–3.5 pounds; also called a broiler

Gallus domesticus—the domestic chicken

Gallus gallus—the Red Jungle Fowl

Genus—a group of closely related animals or birds that differ very slightly from one another (as Gallus gallus from Gallus domesticus)

Gizzard—the tough internal organ where food is macerated

Grade—to sort according to size and quality

Grit—pebbles, sand, or a commercial "grit" product ingested by a chicken and used by the gizzard to grind food

Hackles—a rooster's neck feathers; sometimes collectively called his cape

Hatch—a group of chicks who emerge from their shells at about the same time

Hatching eggs—fertilized eggs carefully stored in a manner that doesn't inhibit their ability to hatch until placed in an incubator or under a hen

Hatchability—the state of being capable of hatching

Hen—a female chicken at least one year old

Hen-feathered rooster—a male chicken having rounded (not pointed) sex feathers

Heritage Chicken—the meat of antique breeds, when the birds are raised to American Livestock Breeds Conservancy standards (see chapter 3)

Humidity—the amount of moisture in the air

Hybrid—A chicken whose parents were two different breeds; when bred together, hybrids don't produce offspring with their own characteristics

Inbred chicken—the offspring of closely related parents; inbreeding is a valuable tool by which breeders set certain characteristics within their chosen bloodlines

Incubation—the act of hatching eggs

Keel—a chicken's breastbone

Large chicken, large fowl—the original, "normal-sized" chicken (as opposed to bantams), sometimes erroneously called standard chickens

Layer, laying hen—a chicken kept for egg production

Litter—the thick layer of cushiony, absorbent material used to line laying nests and poultry building floors; also called bedding

Mate—to pair a male and female animal or bird for breeding purposes

Meat breed chicken—one developed for quick growth and heavy muscling

Molt—the annual shedding and renewal of plumage

Muff—a grouping of feathers bristling out from the sides of bearded breeds' faces; also called whiskers

Nest—a dark, secluded place where a hen feels it's safe to lay her eggs

Nesting box—man-made cubicles placed in hen houses so hens can lay their eggs away from the main flow of traffic

Nest egg—an artificial egg placed in a nest to encourage hens to lay there

Nest run—ungraded eggs

Oviduct—the tube-like internal organ of female birds through which a passing egg is encased in albumen, shell membranes, and shell

Pastured poultry—fowl raised in a pasture setting and housed in moveable shelters

Pasty butt—diarrhea stuck to a chick or older chicken's vent area

Pecking order—the social order of chickens

Perch—(noun) the place, usually elevated rails, where chickens sleep at night; also called a roost; (verb) the act of resting on a roost

Pick out—vent damage caused by other chickens' pecking

Pin bones—pubic bones; two sharp, skinny bones ending near the vent

Pinfeathers—the tips of newly emerging feathers

Plumage—feathers

Pip—(verb) the act of an emerging chick breaking a hole in her shell as part of the hatching process; (noun) the hole a hatching chick makes

Pullet—a female chicken less than one year of age

Purebred—a chicken whose parents are both the same breed

Roaster—a three- to five-month-old meat chicken of either sex weighing 4–6 pounds

Roo—an affection-laced word for rooster

Roost—(noun) the place, usually elevated rails, where chickens sleep at night, also called a perch; (verb) the act of perching

Rooster—a male chicken at least one year of age; also called a cock

Saddle—collectively, the feathers on a rooster's back, just before the tail

Scratch—(verb) the act of scratching the ground in search of food; (noun) any grain product fed to chickens

Set—the act of allowing or encouraging a broody hen to incubate eggs

Sex-link breeds—breeds in which male and female chicks hatch in different colors or patterns, thus making accurate sexing immediately possible

Sexed chicks—all cockerels or all pullets separated by sex

Sex feathers—rounded hackle, saddle, and tail feathers on a hen; pointed hackle, saddle, and tail feathers on a rooster

Shank—the lower part of a chicken's leg between its claw and first joint

Sickle feather—a long, curved rooster tail feather

Shell membranes—two thin membranes immediately inside of an egg shell

Spent hen—a worn-out hen no longer laying well

Spurs—the sharp pointed appendages on a rooster's shanks

Standard—a description in word and picture of a breed's ideal specimen

Started pullets—sixteen- to twenty-two-week-old pullets on the brink of laying; usually purchased from a specialist breeder

Starter—commercial feed ration formulated for newly hatched chicks; there are two formulations, regular and broiler chick starters

Starve out—the act of newly hatched chicks refusing to eat

State Egg Statute—A collection of laws regulating the sale of eggs within a state

Stewing hen—a tough, old hen suitable only for pressurized or moist, slow cooking

Straight run—newly hatched unsexed chicks; a package of straight run chicks contains both cockerels and pullets

Strain—a group of animals or birds within a breed or variety, developed by a single breeder or small group of breeders; animals or birds of the same strain are very uniform and usually share common bloodlines

Stub—down on the legs of a supposedly clean-legged chicken

Tin hen—(slang) an artificial incubator

Trio—a rooster and two hens of the same breed and variety

Type—a breed's look; the distinctive size, shape, and appearance that indicate what breed a chicken belongs to

Variety—a subdivision of a breed based on physical characteristics such as color, comb type, or fancy feathering

Vent—the cloaca's outside opening, through which a chicken eliminates and lays eggs

Wattles—two dangles of red flesh drooping down from the outer edges of a chicken's chin

Zoning—Laws regulating land use, including whether or not landowners can keep chickens

Resources

Hatcheries

Belt Hatchery (California)
559-264-2090
www.belthatchery.com
Belt Hatchery offers the usual commercial and backyard breeds as well as a decent number of fancy breeds.

Cackle Hatchery (Missouri)
417-532-4581
www.cacklehatchery.com
Cackle Hatchery sells a huge array of fancy, heirloom, and commercial day-old chicks. Their specialty is Old English Game varieties—dozens—and all are pictured on their colorful website. They also offer several five- or ten-pullet city chicken specials.

C.M. Estes Hatchery (Missouri)
800-345-1420
www.esteshatchery.com
Estes Hatchery sells a fine selection of commercial and fancy day-old chicks, including bantams. Their website also features great information to get you started.

Healthy Chicks and More (Ohio)
513-238-5735
www.healthychicksandmore.com
Healthy Chicks and More offers all the usual breeds and a nice selection of fancy and rare breeds in shipments as small as ten chicks.

Hoffman Hatchery (Pennsylvania)
717-365-3694
www.hoffmanhatchery.com
Hoffman Hatchery sells poultry-keeping supplies, books, and a wide selection of commercial and fancy breeds.

Ideal Poultry (Texas)
254-697-6677
www.ideal-poultry.com
All the usual chickens and a fine selection of fancies and rare breeds are here in the Ideal Poultry catalog.

Murray McMurray Hatchery (Iowa)
800-456-3280
www.mcmurrayhatchery.com
Chicks of every conceivable breed and

size, hatching eggs, equipment, feed (including a line of natural products) and supplements, books and DVDs, even T-shirts and Amish egg baskets—if it has anything to do with chickens, it's in Murray McMurray's great free catalog.

Meyer Hatchery (Pennsylvania)
800-568-9755
www.meyerhatchery.com
Meyer's carries a huge array of chicks as well as hatching eggs and seventeen-week-old layer pullets. In addition, they offer a full line of chicken-keeping and hatching supplies as well as gift items and chicken diapers.

My Pet Chicken (Connecticut)
888-460-1529
www.mypetchicken.com
My Pet Chicken sells an astounding number of breeds in groups as small as three to eight chicks. They also offer a wide variety of hatching eggs, supplies (including chicken diapers), books, and gifts. Visit My Pet Chicken's website to view oodles of great information and to download their free chick care and incubation e-books.

Privett Hatchery (New Mexico)
877-774-8388 (PRIVETT)
www.privetthatchery.com
Privett Hatchery sells a fine selection of commercial, fancy, and rare-breed chicks, including bantams.

Sand Hill Preservation Center
563-246-2299
www.sandhillpreservation.com
Sand Hill Preservation Center is dedicated to the preservation of more than 1,000 varieties of heirloom vegetables, flowers, fruits, grains, and poultry. They sell scores of common, rare, and heritage breeds via their online and (free) print catalogs.

Stromberg's Chicks (Minnesota)
800-720-1134
www.strombergschickens.com
Stromberg's offers a staggering variety of live poultry and hatching eggs, supplies, books, and DVDs. Download an array of free educational PDFs from their website. You can even buy mature chickens online or from Stromberg's free print catalog.

Welp Hatchery (Iowa; also shipping from New Mexico and Minnesota)
800-458-4473
www.welphatchery.com
Welp Hatchery sells a wide variety of commercial, fancy, and rare-breed day-old chicks along with the equipment and supplies you'll need to raise them.

Organizations

American Bantam Association (ABA)
973-383-8633
www.bantamclub.com
Founded in 1914, the American Bantam Association promotes the breeding, exhibition, and selling of purebred bantam chickens and ducks. Visit its website to join the ABA, access breed-club and member links, purchase books and other merchandise, and read about the organization's many programs.

American Livestock Breeds Conservancy (ALBC)
919-542-5704
www.albc-usa.org
The American Livestock Breeds Conser-vancy works to protect nearly one hundred breeds of cattle, goats, horses, asses, sheep, swine, and poultry from extinction. Clink on "Heritage Chicken" to access a treasure trove of information about old-time, endangered breeds.

American Pastured Poultry Producers Association (APPPA)

http://apppa.org

The APPPA unites pastured poultry producers and distributes pastured poultry resources to consumers and potential producers. Visit their website to download the APPPA brochure or locate a pastured poultry producer in your locale.

American Poultry Association (APA)

724-729-3459

www.amerpoultryassn.com

Founded in 1873, the American Poultry Association sanctions poultry shows and publishes the APA Standard of Perfection (the rules by which purebred poultry is shown), a yearbook, and a quarterly newsletter. Use the pull-down menus at the APA website to access a variety of avian information; their Health series and "Raising Birds in the City" (find it in the "useful information" menu) are especially well written.

Australian Poultry Group

http://apgroup.co

The Australian Poultry Group's website is a portal to a score of Australian poultry clubs and informational websites. Click on "Our Domains" to find them.

National Center for Appropriate Technology (NCAT)

800-346-9140

www.attra.ncat.org/attra-pub/poultry

For the past twenty-five years, the National Center for Appropriate Technology has served economically disadvantaged people by providing information and access to appropriate technologies that can improve their lives. The ATTRA Project, funded by the US Department of Agriculture, is managed by NCAT. ATTRA provides information and other technical assistance to farmers, ranchers, extension agents, educators, and others involved in sustainable agriculture throughout the United States. Visit NCAT/ATTRA's Sustainable Poultry website to view or download dozens of valuable ATTRA bulletins, or phone to request a free information packet tailored just for you.

National Poultry Improvement Plan (NPIP)

www.aphis.usda.gov/animal_health/animal_dis_spec/poultry

The USDA certifies breeders and hatcheries, ensuring that their chickens (and other poultry) test free of the deadly Pullorum Disease. Read about the National Poultry Improvement Plan and locate participating hatcheries via the NPIP online directory.

Poultry Club of Great Britain (PCGB)

www.poultryclub.org

Founded in 1877, the Poultry Club is Britain's equivalent of the American Poultry Association. Everyone will learn a lot from their information-packed website. Be sure to click on "Breed Gallery" and "Eggs."

Rare Breeds Survival Trust (RBST)

www.rbst.org.uk

The Rare Breeds Survival Trust (the United Kingdom's equivalent of the American Livestock Breeds Conservancy) currently montors thirty-one breeds of rare chickens, including many concurrently tracked in North America by the ALBC.

Society for the Preservation of Poultry Antiquities (SPPA)

www.feathersite.com/Poultry/SPPA/SPPA.html

Fascinated by heirloom chickens? Join the SPPA and help preserve and promote them.

Chicken Supplies

Boxes for Birds
501-588-2283
www.boxesforbirds.com
Boxes for Birds sells safe, sturdy shipping boxes for live chickens.

Chicken Diapers
www.chickendiapers.com
Ruth at Chicken Diapers promotes her website as "a clean place for chickens and humans to sit." She pioneered reusable cloth chicken diapers in 2002 and has sewn many thousands since then. Ruth Cahill Haldeman's chicken diapers are the real McCoy!

ChixChix Chicken Diapers
http://chicchix.webs.com
ChixChix offers an alternate diaper design.

Diagraph Quikspray Blockout Ink
800-233-1456
www.diagraphsnyder.com
QuikSpray Blockout Ink covers unwanted marks on egg cartons so they can be reused. It dries quickly and is oil- and water-resistant. It can be purchased in aerosol cans or in one- or five-gallon pails and comes in the colors of tan and white.

Egganic Industries
800-783-6344
www.henspa.com
Eggantic Industries manufactures neat, prefabricated chicken coops on wheels. Their website is a wonderful source of useful information about all things chicken.

Eggboxes
800-326-6667
http://eggboxes.com
Eggboxes sells a wide array of hatchery and egg-seller supplies, including many types of plain or custom-printed egg cartons.

Eggcartons.com
888-852-5340
www.eggcartons.com
Visit this site to order plain or custom-imprinted paper, foam, and plastic egg cartons at discount prices. Also available are incubators, feeders, waterers, books, and a fine selection of chicken-themed gifts.

Egg Cartons Online
877-454-3447 (EGGS)
www.eggcartonsonline.com
Egg Cartons Online markets a huge line of egg cartons, an egg washing kit, hatching supplies, and books.

EZ Clean Coops
888-442-9326
www.ezcleancoops.com
EZ Clean coops builds a fine line of gorgeous, pre-fabricated, wooden chicken houses. You can even buy them as kits!

Horizon Micro-Environments
800-443-2498
www.hm-e.net
Horizon manufactures a large selection of boxes for shipping live chickens.

Kemp's Koops
888-901-2473
www.poultrysupply.com
Kemp's Koops sells chicken, waterfowl, gamebird, and exotic-avian supplies and incubators.

Omlet
866-6538-872
www.omlet.us
Omlet, a British firm (find its British website at http: www.omlet.co.uk) with

an American presence, manufactures a variety of wonderful, prefabricated chicken houses made of easy-to-clean molded plastics. Their American website features a peerless breed directory (you can even review your favorite breed!) and a chicken-raising guide.

Poultryman's Supply Company
812-603-7722
http://poultrymansupply.com
Poultryman's Supply offers incubators, brooders, waterers and feeders, medications, books, leg bands, egg cartons, and much, much more.

Smith Poultry & Game Bird Supply
913-879-2587
www.poultrysupplies.com
Smith Poultry and Game Bird Supply sells incubation supplies, brooders, nest boxes, netting, leg bands, waterers, feeders, medications, vaccines, vitamins, disinfectants, books, and a lot of other products that chicken raisers of all sorts will appreciate.

University Resources
Major state universities and all state extension services distribute papers and bulletins of interest to chicken owners. Compile an up-to-date free library of chicken materials by printing out papers online or downloading PDF files and bulletins and building the printouts to create your own "everything about chickens" resource book.

University of Arkansas Extension Poultry Web Page
www.aragriculture.org/poultry.htm
Visit to access scores of university bulletins. Be sure to click on "Fun with Incubation" and "Small Flock Information."

University of California Poultry Page
http://animalscience.ucdavis.edu/Avian/pubs.htm
Download hundreds of free University of California/Davis poultry fact sheets, leaflets, booklets, and technical papers.

University of Florida Cooperative Extension
http://edis.ifas.ufl.edu
To access the University of Florida Extension's many great chicken publications, click on "Agriculture" under "Topics," then "Livestock," then "Poultry."

University of Georgia Cooperative Extension
www.caes.uga.edu/extension
The University of Georgia College of Agricultural and Environmental Sciences publishes thousands of online and PDF format documents. To locate chicken titles, click on "Agriculture," and then "Poultry."

University of Illinois Extension Incubation and Embryology Pages
www.urbanext.uiuc.edu/eggs/index.html
Written for upper elementary to high school students and their teachers, this material covers everything from constructing an egg candler to incubation troubleshooting.

Kansas State University Research and Extension
www.ksre.ksu.edu
Selling eggs from your farm? Don't miss Kansas State University Extension's bulletin, "Packing Eggs on the Farm for Direct Sales." Hoping to show your chickens? "Selecting and Preparing Poultry for Exhibition" will show you how. Click on "Publications," then "Livestock," and then "Poultry" to access these and more.

University of Kentucky Collage of Agriculture Agripedia
www.ca.uky.edu/agripedia
To access an array of useful web pages, charts, and bulletins, click "Subject Index," then click "C" (for chicken) and "P" (for poultry).

University of Minnesota Extension Service
www.extension.umn.edu
To access University of Minnesota chicken bulletins, click on "Agriculture," then "Poultry," then "Resources" (it's in the left-hand menu). Their "Home Processing of Poultry" bulletin is outstanding.

Mississippi State University Extension Service
http://msucares.com
Finding chicken resources at the "MSU Cares" site is the essence of simplicity. Click on "Poultry" in the menu.

University of Missouri Extension
http://muextension.missouri.edu
University of Missouri Extension offers an impressive selection of useful chicken bulletins. Access them by clicking "Agriculture," then "Animals," and then "Poultry."

University of Nebraska-Lincoln Institute of Agriculture and Natural Resources
http://ianrpubs.unl.edu
Access a small but very good collection of chicken bulletins by clicking "Poultry" under "Browse Publications."

North Carolina State University 4-H Poultry Page
www.ces.ncsu.edu/depts/poulsci/4h/
If you hatch your own eggs, be sure to click on "Embryology;" it's the best!

Ohio State University Extension
http://extension.osu.edu
Click on "Agriculture," then "Poultry" to access Ohio State University Extension's chicken publications.

Oklahoma State University Extension Fact Sheets
http://osuextra.okstate.edu
Click on "Topical List," then "Animals," and then "Poultry" to access a variety of chicken publications.

Penn State Poultry Extension
www.extension.psu.edu
Don't miss this site—Penn State's chicken resource page links to hundreds of useful chicken-keeping bulletins.

Virginia Cooperative Extension
www.ext.vt.edu
To access the Virginia Cooperative Exten-sion's fine array of chicken bul-letins, click "Animal Agriculture," then "Poultry," and there you are!

West Virginia University Extension Service
http://ext.wvu.edu
Peruse West Virginia University Exten-sion's exceptional chicken resources by clicking on "Agriculture," then "Poultry.

City Chickens Websites
Backyard Chickens
www.backyardchickens.com
This great site features articles, coop plans, a forum, and more.

Chicken Crossing
http://chickencrossing.org
A great site dedicated to city- and pet-chicken keepers that includes a good message board.

The City Chicken

http://thecitychicken.com

This comprehensive site features a chick tractor gallery, great articles, FAQs, oodles of great pictures, and the best roundup of chicken laws online.

C.I.T.Y.
(Chickens in the Yard)

www.salemchickens.com

Salem, Oregon, chicken keepers bring you this informative website about keeping chickens in their area.

Duluth City Chickens

http://duluthcitychickens.org

This site is about city chickens in the far North and has a good FAQ section.

Mad City Chickens

www.madcitychickens.com

Chicken keepers in Madison, Wisconsin, bring you a coop preview and a very helpful FAQ section.

Mad City Chickens at YouTube

www.youtube.com/group/
madcitychickens

The Mad City chicken keepers bring you more than 600 great chicken videos.

Urban Chickens

http://urbanchickens.org

Urban Chickens brings you a blog, a forum, articles, and information on the legalities of keeping chickens in town.

Urban Chickens Network Legal Resource Center

http://wiki.urbanchickens.net

The Urban Chickens Network Legal Resource Center brings you a list of cities where you can legally keep chickens along with each city's specific statutes.

Other Useful Websites

American Holistic Veterinary Medicine Association/AHVMA

http://ahvma.org

Visit the American Holistic Veterinary Medicine Association website to locate a holistic vet for your chickens.

Association of Avian Veterinarians

www.aav.org

When your sick or injured chicken needs a specialist, find one via the Association of Avian Veterinarians website.

Brown Egg, Blue Egg

www.browneggblueegg.com

Brown Egg, Blue Egg belongs to Alan Stanford, PhD, whose tips appear in this book. If you love chickens (click on "Stories") or want to learn more about them (the selection of articles at this site is beyond extensive), don't miss Brown Egg, Blue Egg. It's far and above my favorite chicken website.

Chicken Breeds

www.ansi.okstate.edu/breeds/poultry/
index.htm

Click on "Chickens" to read about and view photos of any breed you can think of and then some.

Chicken Feed: The World of Chickens

www.lionsgrip.com/chickens.html

The Chicken Feed website brings you "sources of natural chicken feed, knowledge about traditional ways of feeding chickens around the world and in old times, and health before profit in raising and feeding chickens."

Chicken Resources on the Web

www.ithaca.edu/staff/jhenderson/
chooks/chlinks.html

John. R. Henderson, a librarian at Ithaca

College and a Lodi, New York, hobby farmer, has created the ultimate poultry-link website. While you're there, click on the "ICYouSee Handy Dandy Chicken Chart" link to visit the Internet's most comprehensive chicken breed chart.

Chickensuit
www.chickenssuit.com
Clothing for chickens—would I kid about this? Visit and watch the movies.

The Coop
www.the-coop.org
Looking for a truly comprehensive resource for chicken keepers? Here it is!

DOM_BIRD
www.afn.org/~poultry
Visit the DOM_BIRD home page where chickens, turkeys, and waterfowl rule. Subscribe to the huge, friendly e-mail group, check out the "Poultry Breed Encyclopedia," and read a collection of interesting articles.

Eggbid
www.eggbid.net
It's the eBay of the poultry world—check it out!

Home Grown Poultry
www.homegrownpoultry.com
Home Grown Poultry is an online magazine featuring a free newsletter, archived articles, a forum, FAQs, and a blog.

FeatherSite
www.feathersite.com
Don't miss FeatherSite—it's amazing! Links lead to every conceivable poultry topic, organization, and business online.

Housechickens
http://groups.yahoo.com/group/housechickens

If you keep pet chickens, in or out of the house, subscribe to housechickens, the friendliest and most fascinating of Yahoo's free chicken-related YahooGroups e-mail lists. Most of the fine folks who contributed tips to the book in your hands are housechickens regulars.

Merck Veterinary Manual
www.merckvetmanual.com
The online version of the Merck Veterinary Manual encompasses more than 12 thousand topics and one thousand illustrations searchable by topic, species, specialty, disease, and keyword. Access is free, compliments of Merck, Inc.

Poultry Health Articles at Shagbark Bantams
http://shagbarkbantams.com/contents.htm
Most of these easy-to-understand health articles were written by Shagbark Bantams owner K.J. Theodore and previously appeared in the Poultry Press. "Stress" and "Hatching" are especially useful. Folks who love their chickens are sure to appreciate, "The Emotional Side of Raising Poultry." This is a must-visit site.

World Egg Throwing Federation
http://swatonvintageday.sslpowered.com
This organization is headquartered in Swanton, England. And yes, it's legit!

Suggested Reading
The following books and periodicals were written for small- and medium-scale chicken keepers, rather than free large-scale meat and egg producers. Most are currently in print; the rest are available through out-of-print booksellers or eBay.

American Bantam Association. *Bantam Standard* (current edition). Augusta, NJ:

American Bantam Association. The ABA *Bantam Standard* classifies and describes the standard physical appearance, coloring, and temperament for all recognized breeds of bantam poultry, including chickens.

American Poultry Association. *American Standard of Perfection* (current edition). Burgettstown, PA: American Poultry Association. First published in 1874 and updated every few years, the APA *Standard of Perfection* (commonly referred to as "the Standard") classifies and describes the standard physical appearance, coloring and temperament for all recognized breeds of poultry, including chickens.

Beck-Chenoweth, Herman. *Free-Range Poultry Production, Processing, and Marketing*. Hartshorn, MO: Back Forty Books, 1997. This information-packed volume is the definitive work on developing a free-range poultry business. It's a complete how-to incorporating construction plans, feed formulas, slaughter information, and marketing solutions. A companion video is available. Buy them (along with a plethora of other poultry titles) at www.back40books.com.

Damerow, Gail. *The Chicken Health Handbook*. North Adams, MA: Storey, 1994. If you own chickens, you need *The Chicken Health Handbook*. Virtually everything you need to know about evaluating flock health and treating your chickens' parasites, ailments, and injuries; incubating and brooding chicks; nutrition; anatomy; and even postmortem examinations is presented in easy-to-comprehend terms and packed into this surprisingly inexpensive, 352-page book. The diagnostic charts in this book are outstanding!

————. *Storey's Guide to Raising Chickens: Care, Feeding, Facilities*. North Adams, MA: Storey, 2010. Gail Damerow knows her chickens. Whether you keep laying hens or raise meat for the freezer, and no matter your level of expertise, this is a book you'll refer to time and time again.

————. *Your Chickens; A Kid's Guide to Raising and Showing*. North Adams, MA: Storey, 1993. It would be easy to say this is *Storey's Guide to Raising Chickens* rewritten for kids, but it's so much more. Great pictures and line drawings, interesting layouts, and lively writing make this a great introductory book for chicken lovers of all ages.

Editors of Hobby Farms Magazine. *Popular Farming: Chickens*. Lexington, KY: BowTie Magazines, 2008. The first of *Hobby Farms*' Popular Farming magabooks™, *Chickens*' 128-pages are jam-packed with informative articles and beautiful photographs.

Ekarius, Carol. *Pocketful of Poultry*. North Adams, MA: Storey Publishing, 2007. Hundreds of chickens, ducks, geese, and turkeys are beautifully illustrated in this compact, 272-page field guide, along with facts on each breed's history, size, features, place of origin, special qualities, and conservation status.

————. *Storey's Illustrated Guide to Poultry Breeds*. North Adams, MA: Storey Publishing, 2007. Here is the definitive guide to poultry breeds, bar none. Carol Ekarius describes more than 120 barnyard fowl, from chickens and turkeys to emus and pheasants, in the 288-pages of this lavishly illustrated book. It's a wishbook for poultry lovers. Check it out!

Feldman, Thea. *Who You Callin' Chicken?* New York: Harry N Abrams, 2003. Written for kids ages 4 to 8, but fun for chicken admirers of all ages, this zany volume spotlights a wide variety of plain and fancy chicken breeds, examining their feathers, life cycle, evolution, and more. Fowl photographer Stephen Green-Armytage's fantastic chicken photos are a sheer delight.

Feltwell, Ray. *Small-Scale Poultry-Keeping; A Guide to Free-range Poultry Production.* London, England: Faber & Faber, 2002. In Britain, backyard hen keeping is a longstanding tradition. Everything from do-it-yourself small-scale poultry housing to feeding, health, and breeding is covered in this cleverly written 196-page handbook.

Foreman, Patricia. *City Chicks: Keeping Micro-flocks of Chickens as Garden Helpers, Compost Makers, Bio-recyclers, and Local Food Producers.* Buena Vista, VA: Good Earth Publications, 2010. If you keep urban chickens, especially if you aspire to sell their eggs, you need this thick (464 pages), fact-filled guide to all things chicken.

Glos, Karma E. "Remedies for Health Problems of the Organic Laying Flock; a Compendium and Workbook of Management, Nutrition, Herbal, and Homeopathic Remedies." Available as a free download at http://kingbirdfarm.com/Layerhealthcompendium.pdf. This 60-page guide to treating chickens using homeopathic, herbal, and additional holistic remedies was made possible by a grant from the Sustainable Agriculture Research and Education (SARE) project. An inexpensive print version is also available directly from the author.

Green-Armytage, Stephen. *Extraordinary Chickens.* New York, NY: Harry L. Abrams, 2000.
Exotic show chickens of all sizes, shapes, and colors parade through these pages in 165 photos, each the work of *LIFE* magazine photographer, Stephen Green-Armytage. Fifty breeds are gorgeously pictured and described.

Hams, Fred. *Old Poultry Breeds.* Buckinghamshire, UK: Shire Books, 2000. *Old Poultry Breeds* (like all Shire handbooks) packs more information into forty pages than most books four times its size! Pictures (one hundred of them, fifty-two in color) and history are its strong points; most every breed we know in America (and a lot we don't) are covered in this neat, small volume.

Kilarski, Barbara. *Keep Chickens! Tending Small Flocks in Cities, Suburbs, and Other Small Places.* North Adam, MA: Storey, 2003. Author Barbara Kilarski explains it all—from determining if chickens are legal in your city or suburb, to locating and raising chicks, to feeding and housing them. The text is peppered with poultry facts and stories about the author's three hens. This gentle, fun, and informative volume is (except for this one, of course) my favorite chicken book of all time.

Lee, Andy and Patricia Foreman. *Chicken Tractor: The Permaculture Guide to Happy Hens and Healthy Soil.* Buena Vista, VA:Good Earth Productions, 1998. Here it is, the book that spawned the chicken tractor phenomenon and it's a good one.

———. *Day Range Poultry: Every Chicken Owner's Guide to Grazing Gardens and Improving Pastures.* Buena Vista,

VA: Good Earth Productions, 2002. On beyond Chicken Tractor; it's an everyday guide to grazing chickens in gardens and on grass.

Michigan State University Extension. "4-H Poultry Fitting and Showmanship Member's Guide." Download this free, thirty-page poultry showmanship guide at http://web1.msue.msu.edu/4h/anisci/4H1520_4-HPoultryF&SMG.pdf

Pangman, Judy. *Chicken Coops; 45 Building Plans for Housing Your Flock*. North Adams, MA: Storey Publishing, 2006. The directions in this book are easy to follow and the coops themselves delight the eye. If your chickens need housing, this is your book!

Rossier, Jay. *Living with Chickens; Everything You Need to Know to Raise Your Own Backyard Flock*. Guilford, CT: Lyons Press, 2002 Detailed how-to advice on housing, hatching, buying, feeding, and butchering chickens combined with dozens of detailed illustrations and outstanding photography make this an especially appealing addition to the chicken keeper's bookshelf.

Magazines

Backyard Poultry
www.backyardpoultrymag.com
800-551-5691
Visit this cool bimonthly publication's Web pages to read scores of articles archived on-site.

Chickens
www.hobbyfarms.com/chickens-magazine/
Chickens is published three times a year by the folks who bring you *Hobby Farms*, *Hobby Farm Home*, and *Urban Farm*. Check it out!

Feather Fancier
http://featherfancier.on.ca
Feather Fancier, published eleven times a year in Canada, caters to breeders and fanciers of purebred poultry of all sorts. It's distributed throughout North America and abroad.

Hobby Farms
www.hobbyfarms.com
Hobby Farms is *the* magazine for rural enthusiasts—hobby farmers, small production farmers and those passionate about the country. This bimonthly magazine is devoted to all aspects of rural life, from small farm equipment, to livestock, to crops.

Hobby Farm Home
www.hobbyfarms.com/hobby-farm-home-portal.aspx
Hobby Farm Home is the home magazine for those truly living in the country. *Hobby Farm Home* highlights farmhouse activities such as cooking, crafting, collecting, pet care, and home arts and skills.

Poultry Press
www.poultrypress.com
Poultry Press, established in 1914, is an information-packed monthly not to be missed.

Urban Farm
www.urbanfarmonline.com
From the editors of *Hobby Farms* and *Hobby Farm Home*, *Urban Farm* magazine reaches out to those in the city and suburbs who want to raise chickens, grow food for themselves, support local agriculture, and live more sustainably.

Photo Credits

The sources for the photographs and illustrations for this volume are listed below. Any images not credited here are copyright or courtesy of John and Sue Weaver. Images ending with "/SS" are from Shutterstock.com. Illustrations by Tom Kimball.

Index

Sue Weaver has written hundreds of articles and nine books about livestock and poultry. She is a contributing editor of *Hobby Farms* magazine and writes the Poultry Profiles column for *Chickens* magazine. Sue lives on a small farm in Mammoth Spring, Arkansas, which she shares with her husband of 37 years, a flock of Classic Cheviot sheep and a mixed herd of goats, horses large and small, a donkey who thinks she's a horse, two llamas, a riding steer, a water buffalo, a pet razorback pig, guinea fowl, and Buckeye chickens.